Aims and Scope

The series "Springer Theses" brings together a selection of the very best Ph.D. theses from around the world and across the physical sciences. Nominated and endorsed by two recognized specialists, each published volume has been selected for its scientific excellence and the high impact of its contents for the pertinent field of research. For greater accessibility to non-specialists, the published versions include an extended introduction, as well as a foreword by the student's supervisor explaining the special relevance of the work for the field. As a whole, the series will provide a valuable resource both for newcomers to the research fields described, and for other scientists seeking detailed background information on special questions. Finally, it provides an accredited documentation of the valuable contributions made by today's younger generation of scientists.

Theses may be nominated for publication in this series by heads of department at internationally leading universities or institutes and should fulfill all of the following criteria

- They must be written in good English.
- The topic should fall within the confines of Chemistry, Physics, Earth Sciences, Engineering and related interdisciplinary fields such as Materials, Nanoscience, Chemical Engineering, Complex Systems and Biophysics.
- The work reported in the thesis must represent a significant scientific advance.
- If the thesis includes previously published material, permission to reproduce this must be gained from the respective copyright holder (a maximum 30% of the thesis should be a verbatim reproduction from the author's previous publications).
- They must have been examined and passed during the 12 months prior to nomination.
- Each thesis should include a foreword by the supervisor outlining the significance of its content.
- The theses should have a clearly defined structure including an introduction accessible to new PhD students and scientists not expert in the relevant field.

Indexed by zbMATH.

More information about this series at http://www.springer.com/series/8790

Arild Velure

A Digital Signal Processor for Particle Detectors

Design, Verification and Testing

Doctoral Thesis accepted by
University of Bergen, Bergen, Norway

Springer

Author
Dr. Arild Velure
CERN
Meyrin, Switzerland

Supervisors
Prof. Kjetil Ullaland
Department of Physics and Technology
University of Bergen
Bergen, Norway

Prof. Dieter Röhrich
Department of Physics and Technology
University of Bergen
Bergen, Norway

ISSN 2190-5053 ISSN 2190-5061 (electronic)
Springer Theses
ISBN 978-3-030-71558-8 ISBN 978-3-030-71559-5 (eBook)
https://doi.org/10.1007/978-3-030-71559-5

This Springer imprint is published by the registered company Springer Nature Switzerland AG
The registered company address is: Gewerbestrasse 11, 6330 Cham, Switzerland

Supervisors' Foreword

This thesis describes a microcircuit, the SAMPA chip, which is presently being used by ALICE (A Large Ion Collider Experiment), which is an experiment dedicated to heavy-ion physics at the Large Hadron Collider (LHC) at CERN. It is designed to study the physics of strongly interacting matter at extreme energy densities, where a phase of matter called quark-gluon plasma forms. In experimental high-energy physics, specialized, custom design circuits are widely used. The detectors are large and often need a high number of channels to be readout, for example, the Time Projection Chamber (TPC) has more than 500,000 charge sensitive pads. Each of these needs a low noise amplifier, pulse shaping filter and analogue-to-digital converter, as well as digital signal processing, all of which would require large circuit boards if implemented with standard components. Even if the functional requirements could be met with such a solution, the physical size, cooling needs due to high power consumption and other practical and performance problems would make them highly unattractive.

At the time of writing, the ALICE experiment is being upgraded to take a much higher Pb-Pb collision rate than previously, and consequently all readout electronics have to be replaced by solutions with significantly enhanced performance. Previously data was readout in frames of about 1000 samples triggered by individual particle collisions. But in the upgraded detector, the collision rate is so high that continuous readout is needed, which results in a much higher data output rate from the front-end chips (SAMPAs) to the back-end electronics. The SAMPA front-end chip is one of the most crucial elements in this upgrade and is therefore essential for ALICE to carry out their foreseen future physics program.

Each SAMPA chip has 32 channels. Every channel contains, among others, a charge sensitive amplifier-shaper combination, a 10-bit analogue-to-digital converter running at 10 MHz and a digital signal processing chain with various options for signal baseline stabilization, hit filtering and hit finding. Global infrastructures are, for instance, an I^2C slow control system to write and read a multitude of configuration parameters, a clock distribution and reset network, and a data packer and serializer which can serve several links of 320 Mbit/s and even daisy chain chips in case of lower hit rates.

The operation of several fabricated chip versions requires a significant hardware and software infrastructure to physically send slow control data and timing (clock, synchronization) to the chip and to receive the multiple fast data streams sent by the chip. Dr. Velure achieved this by connecting the chip to a programmable logic circuit (field programmable gate array, FPGA), by creating suited firmware to operate the FPGA and by developing software to control the FPGA and to analyse and visualize the acquired data. This effort was vital in characterizing the device and ensuring that the manufactured devices were produced without fault, so that sub-par devices would not be installed in the experiment, affecting the quality of physics results for many years to come.

The thesis work was performed at the Department of Physics and Technology, University of Bergen, between 05/2013 and 10/2017, leading to several publications and presentations in international conferences, and was honoured as best PhD thesis by the CERN ALICE collaboration.

Bergen, Norway
January 2021

Prof. Kjetil Ullaland
Prof. Dieter Röhrich

Abstract

The A Large Ion Collider Experiment (ALICE) at the Large Hadron Collider at CERN is upgrading two of its sub-detectors, the Time Projection Chamber and Muon Chambers, with new front-end electronics to handle the expected higher Pb-Pb collision-rates in the next running period (Run 3) foreseen to start in 2021. The higher collision rate requires the detectors to employ a continuous readout of the data from the front-end, in contrast to the previous triggered readout. The devices currently employed for the readout of the detectors can only operate in triggered mode and need to be replaced. A new 32-channel integrated circuit called SAMPA has been designed to match the requirements of both detectors. The SAMPA device contains a charge sensitive amplifier, a pulse shaper and a 10-bit 10 MHz analogue-to-digital converter for each channel and a common digital signal processor part. The digital signal processor provides various signal filtering and conditioning operations to improve on the data compression ability. Acquisition can be done in either triggered or continuous mode and the data is offloaded through 320 Mbps differential serial links, allowing a data throughput of up to 3.2 Gbps.

The first prototype of the SAMPA was delivered in 2014, the second in 2016 and the third was delivered in the end of 2017. The final production run was done in mid-2018 and completed the testing at the end of 2018. FEC production and testing is underway and the MCH and TPC are ready for installation in mid-2019.

The main purpose of this thesis has been to specify, design, test and verify the digital signal processing part of the SAMPA device to encompass the needs of the detectors involved. Innovative solutions have been employed to reduce the bandwidth required by the detectors, as well as adaptations to ease data handling later in the processing chain.

By means of simulations, test procedures, verification methods and applied methods for design of reliable systems, a major part of the work has been on qualifying the design for submission to production. Since the design submission process and the following production time of the device is quite long and as the only means of verifying and reading out data from the analogue front-end and the analogue-to-digital converter is through the digital part of the device, it is of the essence to have a complete functioning prototype of the digital design before submission.

A high-speed data acquisition system was developed to enable test and verification of the produced devices. It has been used in all facets of qualification of the device for use by the detectors.

Three rounds of prototypes have been produced and tested. Only minor modifications to the digital design were added between the second and third prototypes. The final production of about 80,000 devices has been completed with the same design as for the third prototype. No major issues have been found in the final design.

The design and test features implemented in the design have been utilized in the production testing and a final yield of close to 80% have been reached.

Acknowledgements

This work has been carried out within the Microelectronics Research Group and the Experimental Nuclear Physics group at the Department of Physics and Technology, University of Bergen, during the period of May 2013 to May 2017.

I would like to thank my supervisors Kjetil Ullaland and Dieter Röhrich for guidance and support throughout the work of this thesis.

I would additionally like to extend my thanks to the SAMPA team for an excellent collaboration. Bruno Sanches, Hugo Daniel Hernandez Herrera and Alex Kluge for which without, this project would have been pale in comparison. Marco Bregant, Marcelo Gameiro Munhoz, Wilhelmus Van Noije and Armando Ayala Pabon for well-executed project organization. And also a special thanks to the rest of the Brazil team, Heitor Neves, Raul Acosta Hernandez, Tiago Oliveira Weber, Dionisio Carvalho, Danielle Moraes, Stelvio Barboza, Shiroma Montali, Heiner Grover Alarcon Cubas, Arthur Lombardi Campos and Andre Luis Do Couto.

I extend my thanks to the detector groups, which I have interacted with during my work and their help in testing the chip to prove its potential. The TPC group; Kenneth F. Read, Charles L. Britton, LLoyd G. Clonts, Anders Oskarsson, Ulf Mjörnmark, David Silvermyr, Jonathan Adolfsson, Torsten Alt, Sebastian Klewin, Matthias Richter, Christian Lippman. The MCH group; Valerie Chambert, Khaing Mon Mon Tun-Lanoë, Hervé Borel, Christophe Suire, Cristophe Flouzat, Clément de Barbarin, Patrice Russo, Emmanuel Rauly, Gwenaël Brulin and Noël Guillaume. Tonko Ljuibicic at the STAR detector and Alexander Pilyar at the Dubna group.

For introducing me to the wonders of the S-ALTRO, I would like to thank Paul Aspell and Tiina Sirea Naaranoja.

Finally I would like to thank my collaborators and colleagues at the University of Bergen and Oslo whom I have shared a great time with; Ganesh Jagannath Tambave, Johan Alme, Sohail Musa Mahmood, Ketil Røed, Attiq Ur Rehman, Chengxin Zhao, Lars Bratrud, Kristian Smeland Ytre-Hauge, Anders Mikkelsen, Anders Østevik, Kristian Philip Engeseth, Christian Torgersen, Inge Nikolai

Torsvik, Damian Wejnerowski, Håvard Rustad Olsen, Simon Voigt Nesbø, Shiming Yang, Øyvind Lye, Hans Schou, Magnus Pallesen and Kavitha Østgaard.

I will always be indebted to my colleagues Ryan Hannigan and Matteo Lupi for their services rendered in proofreading of this thesis.

Contents

Acronyms

a.u.	Arbitrary Units
ADC	Analogue-to-Digital Converter
ALICE	A Large Ion Collider Experiment
ALTRO	ALICE TPC ReadOut
AMS	Analogue Mixed-Signal
AMS	Austria Micro Systems
API	Application Programming Interface
ARM	Advanced RISC Machine
ASIC	Application Specific Integrated Circuit
ASIM	Atmosphere Space Interaction Monitor
ATCA	Advanced Telecommunications Computing Architecture
ATLAS	A Toroidal LHC ApparatuS
ATPG	Automatic Test Pattern Generation
BC1	Baseline Correction 1
BC2	Baseline Correction 2
BC3	Baseline Correction 3
BERT	Bit Error Rate Tester
BFM	Bus Functional Model
BGA	Ball Grid Array
BIST	Built-In Self-Test
BJT	Bipolar Junction Transistor
BNL	Brookhaven National Laboratory
CCD	Charge-Coupled Device
CDC	Clock Domain Crossing
CERF	CERN-EU High Energy Reference Field
CERN	Conseil Européen pour la Recherche Nucléaire
CMOS	Complementary Metal Oxide Semiconductor
CMS	Compact Muon Solenoid
COTS	Commercial Off The Shelf
C-RAM	Chalcogenide RAM

CROCUS	Cluster ReadOut Unit Concentrator System
CRU	Common Readout Unit
CSA	Charge Sensitive Amplifier
CTP	Central Trigger Processor
DAC	Digital to Analogue Converter
DAQ	Data Acquisition
DCS	Detector Control System
DDL	Detector Data Link
DDR	Double Data Rate
DMR	Dual Modular Redundancy
DNL	Differential Non-Linearity
DRAM	Dynamic Random Access Memory
DS	Dual SAMPA
DSC	Digital Signal Conditioning
DSP	Digital Signal Processor
DUT	Device Under Test
EDAC	Error Correction And Detection
EEPROM	Electrically Erasable Programmable Read-Only Memory
ENC	Equivalent Noise Charge
ENOB	Effective Number of Bits
EPROM	Electrically Programmable Read-Only Memory
FBGA	Fine-Pitch Ball Grid Array
FC	Faraday Cup
FEC	Forward Error Correction
FEC	Front-End Card
FIFO	First-In, First-Out
FIR	Finite Impulse Response
FPD	Fixed Pedestal
FPGA	Field Programmable Gate Array
FRAM	Ferroelectric RAM
FWHM	Full Width at Half Maximum
GBT-SCA	GBT Slow Control ASIC
GBTx	Gigabit Transceiver
GdSP	Gas detector digital Signal Processor
GEM	Gas Electron Multiplier
GPIO	General Purpose Input/Output
GSI	Gesellschaft fur Schwerionenforschnung
HAL	HDL Analysis and Lint
HDL	Hardware Description Language
HEH	High Energy Hadrons
I^2C	Inter-Integrated Circuit
IBM	International Business Machines
IC	Integrated Circuit
IFA	Inductive Fault Analysis
IFE	Institute for Energy Technology

IGBT	Insulated Gate Bipolar Transistor
IIR	Infinite Impulse Response
INL	Integrated Non-Linearity
IP	Intellectual Property
IP2	Internet Protocol
JEDEC	Joint Electron Device Engineering Council
JINR	Joint Institute for Nuclear Research
JTAG	Joint Test Action Group
kRad	kilo Rad (1 Gy = 100 Rad)
LEC	Logic Equivalence Check
LED	Light-emitting diode
LET	Linear Energy Transfer
LFSR	Linear Feedback Shift Register
LHC	Large Hadron Collider
LHCb	Large Hadron Collider beauty
LSB	Least Significant Bit
LUT	Look-Up Table
LVCMOS	Low Voltage Complementary Metal Oxide Semiconductor
LVPECL	Low Voltage Positive Emitter Coupled Logic
LVTTL	Low Voltage TTL
MANAS	Multiplexed ANAlogic Signal processor
MANU	MAnas NUmérique
MARC	Muon Arm Readout Chip
MATS	Modified Algorithm Test Sequence
MBU	Multi-Bit Upset
MCH	Muon Tracking Chamber
MCU	Multi-Cell Upset
MID	Muon Identifier
MOS	Metal Oxide Semiconductor
MPD	Multi Purpose Detector
MRAM	Magnetoresistive RAM
MSB	Most Significant Bit
MTR	Muon Tracking
MWPC	Multi-Wire Proportional Chamber
NICA	Nuclotron-based Ion Collider fAcility
NM	New Monitor
NVSRAM	Non-Volatile SRAM
OCL	Oslo Cyclotron Laboratory
OSI	Open Systems Interconnection model
p.o.t.	Particles On Target
PASA	PreAmplifier and ShAper
PATCH	Protocol for Alice Tracking Chamber
PCA16	Programmable Charge Amplifier
PCB	Printed Circuit Board
PCIe	Peripheral Component Interconnect Express

PCRAM	Phase-Change RAM
PIC	Precision Ionization Chamber
PLC	Precision Long Counter
PLL	Phase Locked Loop
PRBS	Pseudo Random Binary Sequence
PROM	Programmable Read-Only Memory
PTB	Physikalisch-Technische Bundesanstalt
PXI	PCI eXtensions for Instrumentation
QGP	Quark-Gluon Plasma
RadMon	Radiation Monitor
RAM	Random Access Memory
RCU	Readout Control Unit
RHIC	Relativistic Heavy-Ion Collider
RMS	Root Mean Square
ROM	Read-Only Memory
RORC	ReadOut Receiver Card
RTL	Register Transfer Level
S-ALTRO	Super ALTRO
SAMPA	Serialized Analogue-digital Multi Purpose ASIC
SAR	Successive Approximation Register
SCA	Slow Control ASIC
SDRAM	Synchronous DRAM
SEB	Single Event Burnout
SECDED	Single Error Correction, Double Error Detection
SEE	Single Event Effects
SEFI	Single Event Functional Interrupt
SEGR	Single Event Gate Rupture
SEL	Single Event Latch-up
SET	Single Event Transient
SEU	Single Event Upset
SFDR	Spurious Free Dynamic Range
SFP	Small Form-factor Pluggable
SINAD	Signal-to-Noise And Distortion
SLVS	Scalable Low-Voltage Signalling
SOLAR	SAMPA to Optical Link for Alice Readout
SPI	Serial Peripheral Interface
SPS	Super Proton Synchrotron
SRAM	Static Random Access Memory
SSRAM	Synchronous SRAM
STAR	Solenoid Tracker at RHIC
TCI	Trigger Crocus Interface
TCP	Transmission Control Protocol
TCP/IP	Transmission Control Protocol/Internet Protocol
TDI	Test Data Input
TDO	Test Data Output

TFBC	Thin Film Breakdown Counter
TID	Total Ionized Dose
TMR	Triple-Modular Redundancy
TPC	Time Projection Chamber
TQFP	Thin Quad Flat Pack
TSMC	Taiwan Semiconductor Manufacturing Company
TSOP	Thin Small-Outline Package
TSOP1	Thin Small-Outline Package 1
TSOP2	Thin Small-Outline Package 2
UART	Universal Asynchronous Receiver/Transmitter
UDP	User Datagram Protocol
USB	Universal Serial Bus
VFBGA	Very Fine-Pitch Ball Grid Array
VHDL	VHSIC Hardware Description Language
VHSIC	Very High Speed Integrated Circuit
VME	Virtual Machine Environment
VPD	Variable Pedestal
VTRx	Versatile Transceiver
μTCA	Micro Telecommunications Computing Architecture

Chapter 1
Introduction

Custom Application Specific Integrated Circuits (ASICs) are often used in the readout of detectors in high-energy physics experiments, as commercial solutions are commonly not available. By using a custom ASIC the detector groups can record data with the lowest noise, lowest power, the highest fidelity and the highest integration, which is additionally tailored to their specific detector.

Some of the electronics for the ALICE detector at the Large Hadron Collider (LHC) at the European Organization for Nuclear Research (CERN) has been installed since the start-up of LHC in 2008. A longer shutdown period is planned for 2019–2020 where the detector-groups will have the possibility to upgrade the detector and electronics to cope with the higher collision rates planned for the following years.

Two of the detectors, the TPC and the MCH, will use a newly designed front-end chip called SAMPA to be able to read out their detector at higher collision rates. Additionally they will replace much of the readout electronics to adapt to the higher data rates that will be produced. This section gives an introduction to the detectors and their motivation for the upgrade.

1.1 The ALICE Experiment

The LHC is a large circular particle collider located on the Swiss-French border close to Geneva, Switzerland. It lies about 100 m below ground in a tunnel 27 km in circumference. Two adjacent pipes each containing particles travelling in opposite direction can be brought together at four interaction points along the beam line as shown in Fig. 1.1. A Large Ion Collider Experiment (ALICE) [1] is located at one of these interaction points and is a general-purpose heavy-ion experiment focused on studying the physics of strong interaction matter and the Quark-Gluon Plasma (QGP) in lead-lead (Pb–Pb) collisions [1].

A. Velure, *A Digital Signal Processor for Particle Detectors*, Springer Theses,
https://doi.org/10.1007/978-3-030-71559-5_1

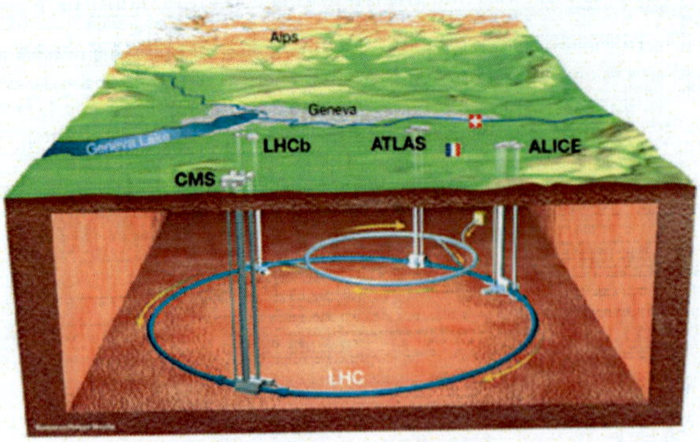

Fig. 1.1 Location of LHC and its four experiments, ALICE, CMS, ATLAS, and LHCb

1.1.1 Physics Goals

There were several epochs in the history of our universe. Strong indications have been found that on the way from the Quark epoch to the Hadron epoch (roughly 1 μs after the Big Bang) a phase of matter called QGP existed. In this kind of matter, the quarks will interact as individual particles instead of being bound together inside of the nucleus. The transition to this de-confinement of quarks happens at a high temperature or at high net-baryon density. This early stage of our universe can be investigated by colliding nuclei at sufficient energies and creating similar conditions. The LHC currently provides the highest energies available in different collision systems with protons and lead ions. The ALICE detector has been built to investigate these type of heavy-ion collisions.

1.1.2 Detector and Sub-detectors

The ALICE detector is constructed of layers upon layers of sub-detectors as shown in Fig. 1.2. Each sub-detector is optimized to study a different aspects of the particles produced in the particle interactions. The two beams enter from the right and left and interact in the centre of the detector.

The functionality of the TPC and MCH will be further described in the next sections as the work described in this thesis is designed for these two sub-detectors.

1.1.2.1 Time Projection Chamber

The Time Projection Chamber (TPC) is one of the main tracking sub-detector of ALICE and is located in the centre of the detector in Fig. 1.2. It is a 90 m^3 gas-filled cylinder with a Multi-Wire Proportional Chamber (MWPC) readout on both

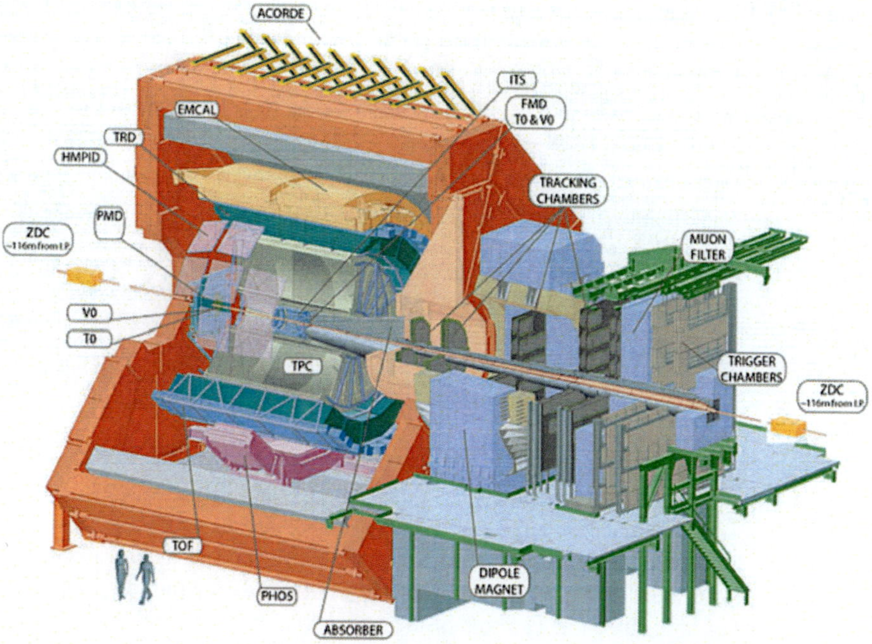

Fig. 1.2 Schematic overview of the ALICE detector showing where the Time Projection Chamber and Muon Tracking Chamber are situated in the detector [1]

end plates, with a total of 557 568 readout pads. A central high voltage electrode located at its axial centre divides the active drift volume into two halves. A schematic overview of the detector can be seen in Fig. 1.3.

The beams colliding at the centre of the cylinder create charged particles, which traverse through the gas, ionizing the gas atoms along the way and creating long tracks. Due to the large magnet that is surrounding the detector, the trace will be bent weaker or stronger in one way or the other depending on the particle's electric charge and momentum. Because of the electric field set up between the centre and the end plates, the liberated electrons created in the ionization will drift at a constant speed towards the readout pads on the end plates. The density of electrons along the track will depend on the momentum and identity of the particle. The end-caps were equipped with MWPCs until 2018; anode wires are strung up close to the end plates, which will create amplification of the signal through the avalanche effect. The positive ion cloud created in the avalanche process induces an image charge on the cathode pads, which is then propagated to the front-end electronics. The signals induced on the arrangement of pads provide a measurement of the track projection onto the end plate. The third coordinate of the track is extracted from the measurement of the drift times of the ionization electrons.

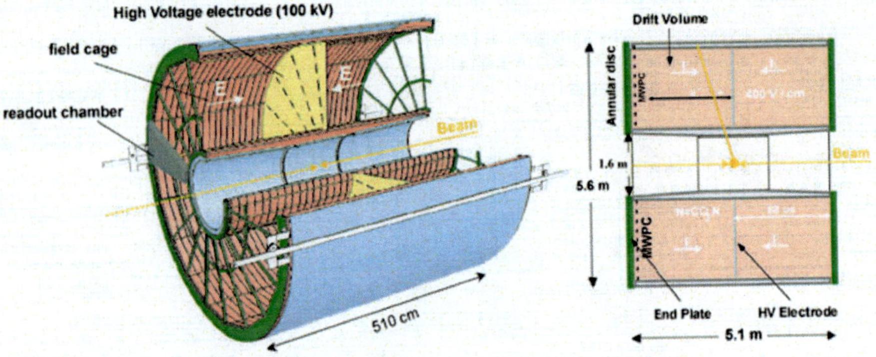

Fig. 1.3 Schematic overview of the time projection chamber

1.1.2.2 Muon Spectrometer

An important piece of evidence for the detection of Quark-Gluon Plasma (QGP) is the yield of quarkonia, which are detected via their decay into dimuons i.e. pairs of muons of opposite sign. The role of the Muon Spectrometer (Dimuon Spectrometer) is to detect muons and measure their momenta from the bending of their tracks in a magnetic field. The Muon Spectrometer is located in the forward region at one side of the barrel, as shown in Fig. 1.2. It consists of an absorber to reduce the large numbers of hadrons, a large dipole magnet to bend the particle tracks, a tracking chamber before, inside, and after the magnet, and a muon filter that protects the two pairs of trigger chamber that follows it. A schematic overview of the spectrometer can be seen in Fig. 1.4.

The tracking system is made of ten planes of MWPCs with cathode pad readout, so-called Cathode Pad Chambers. They are grouped in pairs into five stations for a total surface area of about 100 m^2. With the 1 076 224 readout channels, a spatial resolution of 100 μm can be achieved in the bending plane. The trigger chambers are used to trigger the readout of the tracking chambers [2, 3].

1.1.3 Current Front-End Electronics

The current data readout for most of the detectors in ALICE is trigger based. This means that the faster sub-detectors provide a signal to the slower detectors, like the TPC and MCH, to indicate that they should start the data acquisition and readout. Data from multiple front-end devices are usually then aggregated into readout units, which also control and monitors them. The data from the readout unit is further forwarded to an online system that performs data reconstruction and high level triggering to determine what data to save for further offline analysis.

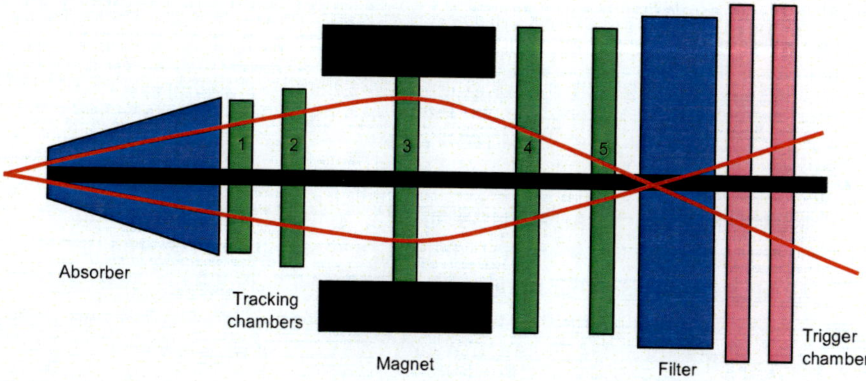

Fig. 1.4 Schematic overview of the Muon Spectrometer showing the absorber, the five tracking chambers, the bending magnet, the muon filter, the trigger chambers, and two particle trajectories

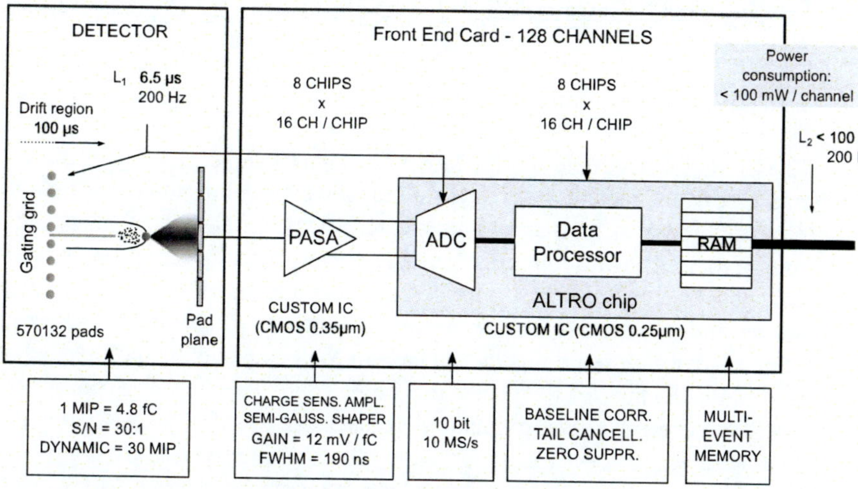

Fig. 1.5 Block diagram of the current TPC front-end card [4]

1.1.3.1 Time Projection Chamber

A single readout channel of the TPC, as shown in Fig. 1.5, is comprised of three basic functional units:

- A charge sensitive amplifier/shaper
- A 10-bit 10 MSPS low power Analogue to Digital Converter (ADC)
- A digital circuit that contains a tail cancellation, baseline subtraction and zero suppression filter, and a multiple-event buffer

The signals from the pads are passed to 4356 front-end cards, located 7 cm away from the pad plane, via flexible Kapton cables. The amplification stage is contained in a separate chip named PASA (PreAmplifier and ShAper) [5], which transforms the pulse from the pad into a differential semi-Gaussian voltage signal on the output. The Analogue to Digital Converter and digital circuits are contained in the ALTRO (ALICE TPC ReadOut) [6] chip. Each chip handles 16 input channels for a total of 128 channels per front-end card, where each channel corresponds to a single pad on the TPC sub-detector.

Each of the front-end cards is connected to a Readout Control Unit (RCU). The RCUs each control between 18 and 25 front-end cards depending on the radial position of the RCU in the TPC barrel. The connectivity between the RCU and the front-end cards is implemented using two branches of a 40-bit wide parallel, multi-drop bus with a bandwidth of 1.6 Gbps per branch. The acquired data is sent to the Data Acquisition (DAQ) system over a 1.28 Gbps optical link (Detector Data Link).

The current system acquires data on a double-trigger scheme, where a trigger is sent to the front-end cards at each interaction, and a second trigger is sent within $\sim$100 μs after the first, indicating whether the data from the current interaction should be kept or discarded.

1.1.3.2 Muon Tracking Chambers

The front-end electronics for the Muon Tracking Chamber is mounted directly on the backside of the readout pads to ensure the lowest level of noise. A single front-end card, referred to as MAnas NUmérique (MANU), handles 64 channels and consists of four MANAS (Multiplexed ANAlogic Signal processor) chips, which themselves contain a Charge Sensitive Amplifier, a deconvolution filter, a semi-Gaussian pulse shaper and a track and hold stage for each of the 16 channels. The 64 analogue outputs are multiplexed together in the MANAS chip and the multiplexed output of two MANAS chips are connected to a single ADC. The ADC is a commercial 12-bit 1 MSPS Successive Approximation Register (SAR) ADC (AD7476) with a serial interface.

A MARC (Muon Arm Readout Chip) chip controls the four MANAS chips and the two ADCs. It also performs zero suppression, pedestal subtraction and encoding of the data, and it additionally handles the communication with the Data Acquisition (DAQ) system through a 4-bit bus. A cluster of Digital Signal Processors (DSPs)[1] called CROCUS (Cluster ReadOut Unit Concentrator System) [7] does the Data Acquisition.

The readout of the current system is triggered by the Muon Trigger chambers located behind the MCH.

[1]ADSP-21160M.

1.2 LHC Run 3 Upgrades

In November 2018, the LHC finished its second running period (Run 2), which started in September 2015. In this period, a peak interaction rate of 7.5 kHz was reached for Pb–Pb collisions with centre-of-mass energy of 6.3 TeV per nucleon pair [8]. As a comparison, in the heavy ion run in 2011 the interaction rate reached was 3 kHz [9].

There is a two year planned shutdown to prepare for Run 3 starting in 2021. Many of the planned measurements for Run 3 will involve complex probes at low transverse momentum, where traditional methods of triggering will not be applicable. Therefore, the ALICE collaboration is planning to upgrade the current detector by enhancing its low-momentum vertexing and tracking capability, and allowing data taking at substantially higher interaction rates.

Run 3 has an expected peak luminosity of $6 \times 10^{27}\,\mathrm{cm}^{-2}\,\mathrm{s}^{-1}$ and interaction rates of about 50 kHz for Pb–Pb collisions [10]. In the proposed plan, the ALICE detector is modified such that all interactions will be inspected. This implies a major upgrade of the TPC detector and a modification of the readout electronics of other detectors to comply with the high readout rate.

1.2.1 Motivation for TPC Upgrade

The drift time of electrons, from centre of the detector to the end plates, in the currently used Ne-CO_2 (90-10) gas in the TPC is $\sim$100 µs. The drift time of positive ions from the amplification region around the MWPC anode wires back to the central electrode is $\sim$180 µs. To avoid any ions drifting back from the amplification region to the drift region, a gating grid is in place, which is enabled after the initial electron drift time, preventing any back drift, but also preventing electrons to pass into the amplification region. This leads to an intrinsic dead time of $\sim$280 µs and a limitation in the maximum interaction rate of 3.5 kHz.

If the gating grid structure were to be permanently disabled and the TPC ran at the targeted interaction rate of 50 kHz (20 µs) for Run 3, then space charge distortions would occur due to the accumulated ions in the drift region, which would render track reconstructions useless. Operation of the TPC at 50 kHz can thus not be accomplished with the current gating scheme [11]. It will therefore be replaced by a multi-stage Gas Electron Multiplier (GEM) system [12]. GEMs have been proven to operate reliably in high-rate applications and provide intrinsic ion blocking capabilities, therefore enabling the TPC to operate in a continuous, ungated readout mode where the readout is dead-time free [13].

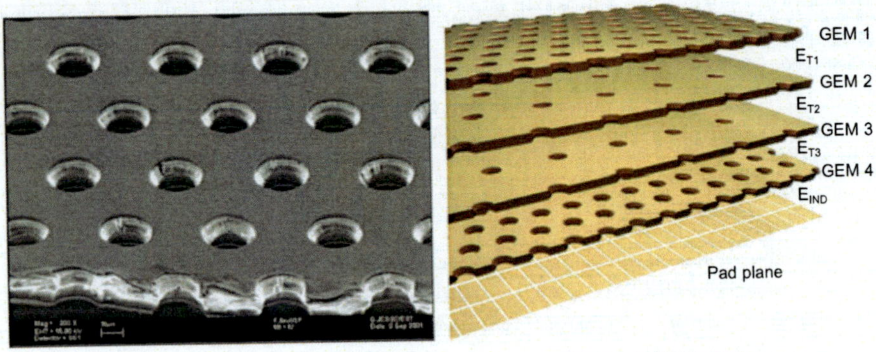

(a) *Close-up of GEM construction* (b) *TPC GEM stackup*

Fig. 1.6 Left image shows an electron microscope picture of a 50 µm thick GEM foil with 70 µm holes and 140 µm spacing [12]. Right picture shows the stackup of four GEM foils, as will be used in the TPC

As the current readout electronics is based around a triggered readout, a full redesign of the complete front-end chain is needed [14]. The new electronics must implement a continuous readout scheme and should be able to handle the resulting higher readout data rate. In addition, it should accommodate both the negative signal polarity of the new GEM detectors and the lower gas gain, which demands a low noise design.

1.2.2 Principal Operation of GEMs

The GEM detector is constructed of a thin foil of Kapton, clad in copper on both sides and chemically pierced with small holes a fraction of a millimetre apart, see Fig. 1.6. A high voltage potential is set up between the two sides of the foil, making large electric fields in the holes. Electrons in the gas entering the hole will create an avalanche of hundreds of electrons. By having several layers of GEMs the number of electrons can be increased even more by providing additional stages of amplification. By offsetting the holes between the layers, the ion backflow can be decreased significantly [12]. The electrons are collected by a suitable device, like a pad plane. A stackup as to be used in the TPC detector can be seen in Fig. 1.6.

1.2.3 Motivation for MCH Upgrade

The triggering source for readout of the Muon Tracking Chamber is the Muon Tracking (MTR), which is currently limited to a trigger rate of 1 kHz. The design read-out rate for Run 3 for MCH has been set to 100 kHz to have a safety margin above the expected rate of 50 kHz. It was found that it is not feasible to upgrade the MTR for these rates and the data from the MTR will instead be used offline for hadron rejection.

Lacking a suitable trigger source, the MCH will instead read out data continuously, requiring a redesign of the front-end card and readout electronics.

1.3 Primary Objective and Main Contributions

The SAMPA project started in the winter of 2012 with the early design of the front-end, and later, the early prototype of the ADC, by Hugo Hernandez of the Electrical Engineering-Polytechnical School, University of São Paulo, Brazil. Heitor Guzzo Neves started initial development for the first prototype of the digital design for the SAMPA ASIC, primarily based on the Super ALTRO (S-ALTRO), in early 2013. He was later assisted by Bruno Sanchez, both of the Electrical Engineering-Polytechnical School, University of São Paulo, Brazil. I joined the project in mid-2013 for firmware verification and testing, as described in Chap. 4. Together with Alex Kluge, the digital and system level specifications for a 32 channel mixed-signal front-end chip meeting the requirements for both TPC and MCH detectors, together with the architectural design of the digital design was defined in late 2013, as presented in Sect. 2.4 and Chap. 3 respectively. Heitor left the project in late 2014, and I took over the responsibility of the development of the design. Optimization of the S-ALTRO filters for more flexibility and higher reliability in the SAMPA, in combination with the testbenches needed for their verification, was done in collaboration with Bruno. An exception is the Baseline Correction 3 (BC3) filter which was developed by Konstantin Münning of the Universität Bonn in collaboration with Bruno. Coding of the base I^2C module used by the design was done by Sandro Bonacini from CERN. However, the design was modified for readability and full test framework was added. Furthermore, my development of an efficient readout scheme to meet bandwidth requirements for the system is presented in Sect. 2.4.3, my design and implementation of a high speed data acquisition and analysis system for use in characterization of the SAMPA ASIC presented in Sect. 4.2, and the experimental characterization of the SAMPA ASIC is presented in Sect. 4.3.

Overall, the development of the SAMPA ASIC has been orchestrated by the Electrical Engineering-Polytechnical School, University of São Paulo, Brazil.

References

1. The ALICE Collaboration, "The ALICE Experiment at the CERN LHC," *Journal of Instrumentation*, vol. 3, no. 08, p. S08002, Aug 2008.
2. The ALICE Collaboration, "ALICE Dimuon forward spectrometer," CERN, Tech. Design Rep. CERN-LHCC-1999-022, 1999. [Online]. Available: http://cds.cern.ch/record/401974.

3. The ALICE Collaboration, "ALICE Dimuon forward spectrometer," CERN, Tech. Design Rep. Add. CERN-LHCC-2000-046, 2000. [Online]. Available: http://cds.cern.ch/record/494265.
4. L. Musa, J. Baechler, N. Bialas, R. Bramm, R. Campagnolo, C. Engster, F. Formenti, U. Bonnes, R. E. Bosch, U. Frankenfeld, P. Glassel, C. Gonzales, H.-A. Gustafsson, A. Jimenez, A. Junique, J. Lien, V. Lindenstruth, B. Mota, P. Braun-Munzinger, H. Oeschler, L. Osterman, R. Renfordt, G. Ruschmann, D. Rohrich, H.-R. Schmidt, J. Stachel, A.-K. Soltveit, and K. Ullaland, "The ALICE TPC front-end electronics," in *2003 IEEE Nuclear Science Symposium. Conference Record*, vol. 5, Oct 2003, pp. 3647–3651 Vol.5.
5. H. Soltveit, J. Stachel, P. Braun-Munzinger, L. Musa, H. Gustafsson, U. Bonnes, H. Oeschler, L. Osterman, and S. Lang, "The PreAmplifier ShAper for the ALICE TPC detector," *Nuclear Instruments and Methods in Physics Research Section A: Accelerators, Spectrometers, Detectors and Associated Equipment*, vol. 676, pp. 106–119, Jun 2012.
6. R. Bosch, A. J. de Parga, B. Mota, and L. Musa, "The ALTRO chip: A 16-channel A/D converter and digital processor for gas detectors," *IEEE Transactions on Nuclear Science*, vol. 50, no. 6, pp. 2460–2469, Dec 2003.
7. The ALICE Collaboration, "The Electronics of ALICE Dimuon Tracking Chambers," CERN, Tech. Rep. ALICE-INT-2004-026, 2004.
8. J. Jowett, "LHC-HI: Summary of Run 2, perspectives for Run 3 and Run 4," ALICE Week, 2018. [Online]. Available: https://indico.cern.ch/event/773102/contributions/3212454/attachments/1768674/2872785/ALICE_Week_Jowett_10Dec2018.pdf.
9. K. Šafařík, "Overview of recent ALICE results," *Nuclear Physics A*, vol. 904-905, pp. 27c–34c, may 2013.
10. B. Abelev and The ALICE Collaboration, "Upgrade of the ALICE experiment: Letter of intent," *Journal of Physics G: Nuclear and Particle Physics*, vol. 41, no. 8, p. 087001, Jul 2014.
11. The ALICE Collaboration, "Upgrade of the ALICE Time Projection Chamber," CERN, Tech. Design Rep. CERN-LHCC-2013-020, ALICE-TDR-016, 2013. [Online]. Available: http://cds.cern.ch/record/1622286.
12. F. Sauli, "The Gas Electron Multiplier (GEM): Operating principles and applications," *Nuclear Instruments and Methods in Physics Research Section A: Accelerators, Spectrometers, Detectors and Associated Equipment*, vol. 805, pp. 2–24, 1 2016.
13. C. Lippmann, "A continuous read-out TPC for the ALICE upgrade," *Nuclear Instruments and Methods in Physics Research Section A: Accelerators, Spectrometers, Detectors and Associated Equipment*, vol. 824, pp. 543–547, 2016, frontier Detectors for Frontier Physics: Proceedings of the 13th Pisa Meeting on Advanced Detectors.
14. The ALICE Collaboration, "Upgrade of the ALICE read-out and trigger system," CERN, Tech. Design Rep. CERN-LHCC-2013-019, ALICE-TDR-015, 2013. [Online]. Available: http://cds.cern.ch/record/1603472.

Chapter 2
Front-End Electronics

The currently installed front-end electronics for both TPC and MCH is the same as what was installed during the commissioning of ALICE in 2008. The planning and ASIC designs, however, started long before this. A lot has improved in the electronic industry since then. We have devices with lower power, more accurate and efficient ADCs, and faster digital electronics.

This section reviews the existing electronics and presents the design requirements for the new system. The devices are compared to the requirements and a digital specification is formulated.

2.1 Heritage

The design and specifications for the SAMPA chip builds on the existing front-end electronics of the TPC, which uses the PASA [1] analogue front-end chip and the ALTRO [2–4] sampling and digital signal conditioning chip. These chips were purposely designed for use in the ALICE TPC. To reduce complexity, and improve noise and signal integrity, the PASA chip was built with a fixed polarity, gain and shaping time, even though this somewhat constrained the detector applications that the chip could be used for. At a later stage the Programmable Charge Amplifier (PCA16) [5] was designed at CERN, which provided programmability for all the settings.

Further development was driven by the needs of the future Linear Collider (ILC/CLIC) to have a compact solution where the analogue front-end was integrated with the digital. This ended with combining the features of the PCA16 with the ALTRO and creating the Super ALTRO (S-ALTRO) [6–9]. The design kept the programmability of the PCA16 and implemented a digital design that more or less

A. Velure, *A Digital Signal Processor for Particle Detectors*, Springer Theses,
https://doi.org/10.1007/978-3-030-71559-5_2

retained the design of the ALTRO intact, with some minor improvements to the digital signal conditioning.

For the third run period of the LHC, the ATLAS CMS[1] GEM detector is also planning to update its front-end electronics. Early in the research phase for the upgrade of the CMS GEM detector, a new front-end chip called the Gas detector digital Signal Processor (GdSP) [10] was proposed and evaluated. It was planned to consist of 64 or 128 channels, with accompanying ADCs, and with a reworked version of the baseline filtering chain from the S-ALTRO in addition to a new digital readout compatible with the Gigabit Transceiver (GBTx) chip [11]. Preliminary analysis work and improvements were done on the filtering and compression architecture before the project decided to use a comparator instead of an ADC, and also to only read out time over threshold measurements, making the baseline filtering capability unneeded.

The SAMPA project bases its analogue front-end on the design of the PASA chip and the filter and zero suppression compression from all of the previous developments.

2.1.1 PASA

The PASA ASIC is a 16-channel preamplifier and shaper chip for use with gaseous detectors. The device was originally custom designed for the ALICE TPC detector [1] and is fabricated in an Austria Micro Systems (AMS) 0.35 µm CMOS technology. The manufactured design has a shaping time of 190 ns, a conversion gain of 12.8 mV/fC @ 12 pF, a noise of 385 e @ 12 pF and it accepts positive polarity input signals. Together with the ALTRO chip, it was measured to have an overall system noise of 710 e when installed in the TPC detector. Table 2.1 lists the tested specifications for the design.

The block diagram of the processing chain for a single channel is shown in Fig. 2.1. Each channel consists of a positive polarity Charge Sensitive Amplifier (CSA) with a capacitive feedback C_f and a resistive feedback R_f (M_f) connected in parallel. The input charge is integrated through the feedback capacitor C_f, while the parallel resistor R_f is used to discharge the capacitor leading the signal to the baseline level with an exponential tail. Following the CSA is a Pole-Zero Cancellation network to remove undershoot in the tail of the signal. The signal is then further amplified and shaped through a fourth-order semi-Gaussian shaper. The final stage adapts the signal levels to match the input of the ALTRO ADC.

[1] A Toroidal LHC ApparatuS Compact Muon Solenoid.

Table 2.1 Summary of tested specifications for the PASA [1]

Parameter	Tested value
Voltage supply	3.3 V
Technology	AMS 0.35 µm
Channels per chip	16
Polarity	Positive
Detector capacitance (C_d)	12.5 pF
Peaking time (t_s)	190 ns
Noise (ENC)	385 *e* @ 12 pF
Sensitivity (gain)	12.8 mV/fC @ 12 pF
Non-Linearity	0.2%
Crosstalk	<0.1%
Power consumption (per ch)	11.67 mW
Die size	18 mm^2

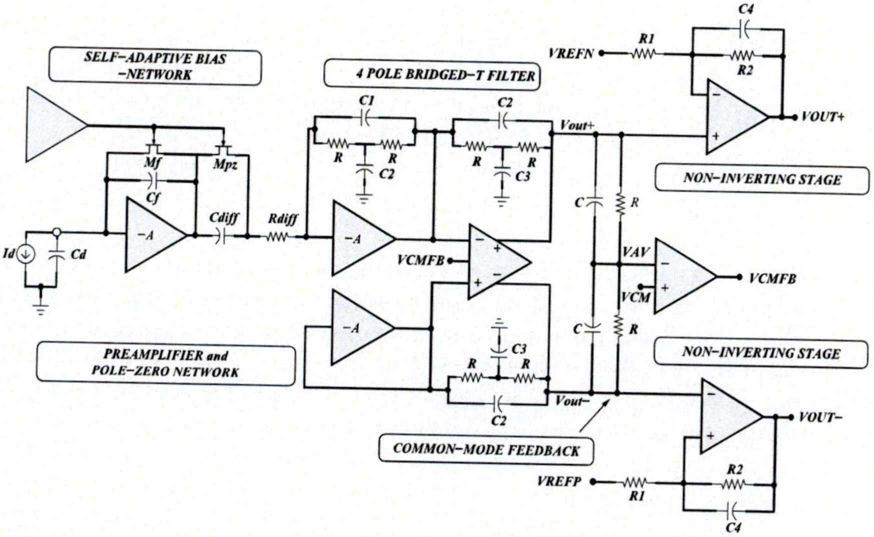

Fig. 2.1 A simplified block diagram of the PASA signal processing chain [1]

2.1.2 ALTRO/S-ALTRO

The ALTRO chip [2–4, 12] is a mixed signal analogue-digital integrated circuit dedicated to the digitization and processing of gaseous detector signals. It was originally designed for use in the ALICE TPC detector, though it has also been used in other detectors. The ADC used for the ALTRO is a commercial pipelined ADC provided by ST Microelectronics, while the S-ALTRO uses a custom designed pipelined ADC [9].

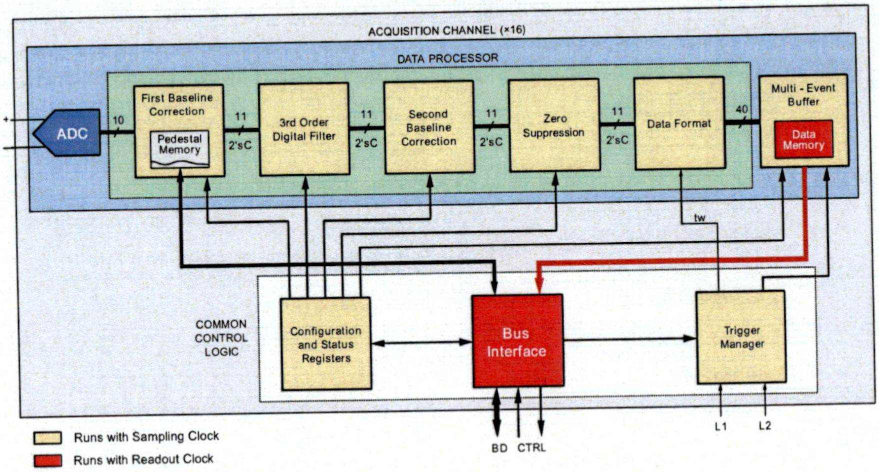

Fig. 2.2 A simplified block diagram of the ALTRO signal processing chain [12]

Both devices are designed to process trains of pulses riding on a semi-stable baseline. A block diagram can be seen in Fig. 2.2 and the specifications are given in Table 2.2. They operate in a triggered fashion where one trigger initiates the acquisition whereupon a predefined number of samples are acquired and temporarily stored in a memory. If a second acceptance-trigger is received, the data is held for later readout, else it is overwritten when the next acquisition is started. The data memory can temporarily store up to eight of these acquisitions.

The data processing blocks of the design implement several stages of signal conditioning to remove baseline variations and to improve the quality of the output signal. The task of the first baseline correction is to remove systematic effects or low-frequency perturbations of the signal baseline. The second section aids in suppressing the tail of the pulse to help with pile-up effects or undershoots. The subsequent block implements a moving average filter to remove non-systematic perturbations of the baseline. After the filtering, the signal baseline should be constant to within one ADC count, which allows an efficient compression using zero suppression technique. The zero suppression compression discards all samples below a given threshold. The cluster of samples above the threshold is assigned a time-stamp and a size information for later reconstruction. The data is further packed into 40-bit words in a back-linked list fashion for subsequent transmission to the readout unit. The data is transmitted over a 40-bit wide bus at up to 60 MHz for a total bandwidth of 2.4 Gbps.

The main drawback of the ALTRO and S-ALTRO chips are that they do not support continuous readout. The designs are constructed so that they can either acquire data to the local buffer or transmit the data off-chip, but not both at the same time. If both were to happen at the same time, the switching noise generated by the 40-bit single-ended parallel-bus readout would swamp the sensitive analogue inputs in noise.

Table 2.2 Summary of tested specifications for the ALTRO [2] and S-ALTRO [6]

Parameter	ALTRO	S-ALTRO
Voltage supply	2.5 V	2.5 V
Technology	ST 0.25 μm	IBM 0.13 μm
Channels per chip	16	16
ADC effective input range	2 Vpp	2 Vpp
ADC resolution	10-bit	10-bit
ADC max sampling frequency	25 MSps	40 MSps
Integrated Non-Linearity	<0.8 LSB	0.71 LSB
Differential Non-Linearity	<0.2 LSB	0.58 LSB
Spurious Free Dynamic Range	78 dBc	N/A
Signal-to-Noise And Distortion	60 dB	56 dB
Effective Number Of Bits	9.7 bit	9.07 bit
Crosstalk	0.005%	0.7%
Noise	0.35 LSB	547 *e*
Power consumption (per ch)	20 mW	47 mW
Die size	64 mm^2	50 mm^2

ALTRO ADC parameters obtained at 10 MSps sampling speed
S-ALTRO ADC parameters obtained at 40 MSps sampling speed

2.2 Common Readout Electronics

Some common projects have been developed in the ALICE and CERN community to increase development efficiency and reduce costs. The Gigabit Transceiver (GBTx) [11] is one of these and provides radiation hardened time-division-multiplexing of multiple serial links to and from a high-speed serial link. Combined with a radiation hardened electrical-to-optical transceiver like the Versatile Transceiver (VTRx) [13], it is possible to communicate over optical links with off-detector electronics that are positioned away from the radiation field of the detector, avoiding the need to design the controller with radiation effect mitigation in mind. This lowers both the development time and cost of a new front-end system. The Common Readout Unit (CRU) [14] is such an off-detector controller, which is being designed as a common project for the Run 3 upgrade of the ALICE detector. The CRU acts as the physical and logical interface to the ALICE online farm, to the Detector Control System (DCS) and to the trigger system for the front-end electronics.

For slow-control and monitoring of front-end cards, a device called the GBT Slow Control ASIC (GBT-SCA) [15] is available. It provides commonly used protocols like SPI, I^{2}C as well as General Purpose Input/Output (GPIO) and ADCs. The GBT-SCA can be used in conjunction with the GBTx to let off-detector electronics easily interface with devices on the front-end cards, without the need for the detectors to design their own front-end board-controller.

2.3 Design Requirements

The main analogue design requirements for both detectors are presented in Table 2.3. They remain mostly the same as before the upgrade and are primarily based on the properties of the detectors and the signal quality required for recreating particle tracks in the offline analysis with sufficient precision. For further information on the analogue design and requirements see [16, 17] respectively. The digital requirements will be further discussed in this section.

Table 2.3 Design specifications for the SAMPA ASIC [17], taken from simulations

Specification	TPC	Muon Chambers
Voltage supply	1.25 V	1.25 V
Polarity	Negative	Positive
Detector capacitance (C_d)	18.5 pF	40 pF–80 pF
Peaking time (t_s)	160 ns	300 ns
Noise (Equivalent Noise Charge (ENC))	<600 *e**	<950 *e*@40 pF*
		<1600 *e*@80 pF*
Linear Range	100 fC / 67 fC	500 fC
Sensitivity	20 mV/fC / 30 mV/fC	4 mV/fC
Return to baseline time	<288 ns	<541 ns
Non-Linearity (CSA + Shaper)	<1%	<1%
Crosstalk	<0.3%	<0.2%
ADC effective input range	2 Vpp	2 Vpp
ADC resolution	10-bit	10-bit
Sampling Frequency	10 MSps or 20 MSps	10 MSps
Integrated Non-Linearity (ADC)	<0.65 LSB	<0.65 LSB
Differential Non-Linearity (ADC)	<0.6 LSB	<0.6 LSB
Spurious Free Dynamic Range (ADC) **	68 dBc	68 dBc
Signal-to-Noise And Distortion (ADC) **	57 dB	57 dB
Effective Number Of Bits (ADC)	>9.2 bit	>9.2 bit
Power consumption (per ch)		
ADC	2 mW (4 mW)	2 mW (4 mW)
CSA + Shaper	6 mW	6 mW
Channels per chip	32	32

* $R_{esd} = 70\,\Omega$
** @ 0.5 MHz, 10 MSps

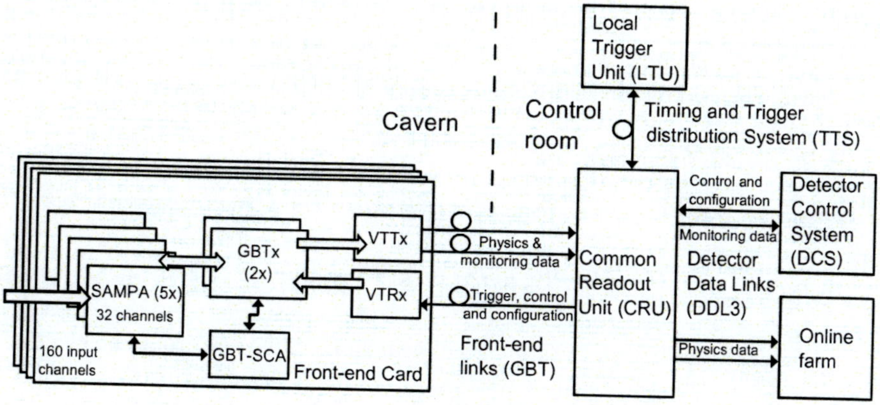

Fig. 2.3 Overview of the TPC readout architecture for Run 3

2.3.1 TPC Readout Electronics

The TPC plans to maintain the existing form factor and placement of the front-end cards in the detector for Run 3. An overview of the readout architecture can be seen in Fig. 2.3. Each front-end card will have five SAMPAs to collect signals from 160 input channels. Flexible Kapton cables connect the front-end card to the detector pads. The digitized data will be sent over high-speed optical links to CRUs situated in the control room, outside the radiation environment of the detector cavern. Each CRU interfaces to, on average, nine front-end cards each. The CRU replaces the previously used RCU that controlled and aggregated the data from multiple front-end cards.

An option is to place the CRU in the detector cavern, in close proximity to the front-end cards. The CRU would connect to the front-end cards through copper cables, instead of optical links, and communicate with the online system through faster optical 10 GbE Detector Data Link 3 (DDL3) links. As the GBTx is limited to a data uplink speed of 4.48 Gbps in the wide bus mode, only half the number of optical links would be needed, reducing the material cost for the installation.

This solution would though restrict the design to using low performance radiation tolerant Field Programmable Gate Arrays (FPGAs). There is additionally the consideration of limited access to the detector cavern during LHC operation, difficult installation and maintenance, extra design work to make the design radiation tolerant, and the relatively costly radiation-verification campaign of the electronic components.

Locating the CRU in the control room outside of the radiation area thus presents a cleaner and more robust solution and enables the use of commercial off-the-shelf hardware.

The CRU controls the configuration, readout and monitoring of the front-end cards and the trigger handling. When the data are forwarded from the CRU to the online system, the individual data fragments are re-ordered according to geometrical position of the originating pad in the detector. In addition, a cluster finder algorithm will be implemented on the CRU that locates clusters that cross more than one pad, and helps in compressing the data further by only forwarding the collected charge of the cluster instead of the individual charge per pad. This reduces the amount of data by a factor 5 to 7 [18].

2.3.2 *MCH Readout Electronics*

As the MCH detector itself will not be upgraded, the new front-end electronics will instead need to compatible with the old form factors and connections. An overview of the readout architecture can be seen in Fig. 2.4.

New Dual SAMPA (DS) boards containing two SAMPA chips each will replace the previous MANU boards, each connecting to 64 detector pads. The configuration and the readout of all the SAMPA chips is ensured by about 650 SAMPA to Optical Link for Alice Readout (SOLAR) concentration cards, replacing the previous CROCUS cards. The SOLAR boards further connect optically to the CRUs over optical links. The CRU controls the configuration, readout and monitoring of the front-end cards, in addition to doing the trigger handling.

Each SOLAR card can be connected to up to 40 DS boards over 6 m flexible Kapton cables. The DS boards are grouped in clusters of up to five boards per cable. Each SAMPA chip can be configured individually, while up to 10 SAMPA chips share the trigger and synchronization signals. On each DS board, the two SAMPA chips share a common 80 Mbps uplink for data to the SOLAR board.

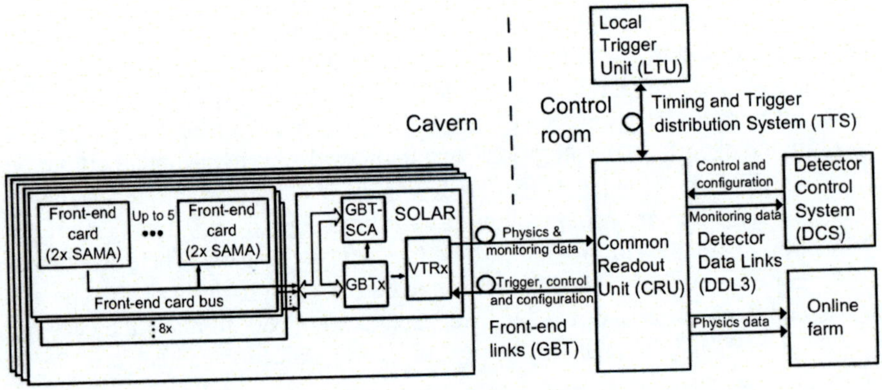

Fig. 2.4 Overview of the MCH readout architecture for Run 3

2.3.3 Common Requirements

As both the TPC and MCH operate in the same ALICE environment and they both plan to utilize the GBTx and GBT-SCA devices in conjunction with the CRU readout electronics, many of their requirements for the SAMPA design will be the same. This section presents the common requirements and the next sections present the TPC and MCH specific requirements respectively.

2.3.3.1 Technology

CERN has experience in previous designs of low noise radiation tolerant devices in the IBM 130 nm technology, like in the design of the predecessor chip S-ALTRO. However, the 130 nm technology is becoming an ageing process and as IBM decided to sell off its foundry business to GlobalFoundries, the future of the process is somewhat uncertain. The design will instead be done in a TSMC process. TSMC provide a 130 nm technology and a 65 nm technology in their advanced process node. The 130 nm technology was chosen for this design as there are generally more problems designing analogue circuits in lower process nodes, e.g. problems with higher noise, lower voltage headroom, transistor matching, radiation sensitivity, etc. Even though there are benefits of lower size, higher clock speeds and lower power. It has though been later shown that the 65 nm process node also is suitable for design of sensitive front-end electronics [19, 20].

2.3.3.2 Noise

The simulated TPC SAMPA noise requirements are the same as the simulated noise for the PASA [18], though at a higher gain which gives some extra headroom. The TPC aims at retaining the same system noise of 670 e as of today. The challenge is that the digital and ADC now are integrated together with the sensitive analogue electronics. Additionally the digital section is constantly processing and sending data off-chip, while the analogue section is acquiring data. The S-ALTRO proved that it is possible to integrate digital and analogue and still reach 550 e [6], measured un-bonded and without input capacitance, so the requirement is viable. The use of low signal swing differential lines for the data transmission and taking precautions in shielding the interface between the digital and analogue domain should satisfy the second concern.

In the case of the MCH, due to their low gain, the noise will be dominated by the resolution and accuracy of the ADC. The noise requirement of 1000 e is equal to 0.64 mV at 4 mV/fC, which is 0.3 LSB. Depending on if the baseline is in the transition part between two codes or between two transition points, the noise will vary between no noise and 0.5 LSB noise.

2.3.3.3 Data Compression Using Zero Suppression

Compression of the sampled data is done by suppressing values below a certain threshold, leaving only cluster data. The remaining data is then run-length encoded. This method is highly dependent on having a stable baseline to achieve good compression and minimal loss of information, so different forms of digital signal conditioning are needed, depending on the detector. The compression is a lossy technique as small signals or parts of signals might be lost. The achievable compression ratio for gas-based detectors with a shaping front-end is generally inversely proportional to the detector occupancy [21].

2.3.3.4 Baseline Correction

Due to manufacturing variations, there will be channel-to-channel and chip-to-chip variations in the offset value of the analogue front-end outputs. This is commonly referred to as a channels baseline or pedestal value. To avoid that this value is below the lower operating range of the ADC, and to be able to include slight undershoots in the tail of the pulse in the digitized data, a small positive offset is added in the analogue front-end before the digital conversion. As the offset will vary from channel to channel, it should be subtracted in the digital domain so that the pulses can be accurately reconstructed later in the analysis.

The distribution of the baseline for each channel in the detector is determined in a dedicated data-taking run, where no collisions occur, so that only the fixed baseline and the superimposed noise can be measured. The average value of the distribution is used for the baseline subtraction and the deviation is used to set the zero suppression level.

2.3.3.5 Power Consumption

Both the TPC and MCH are planning to reuse the existing cooling system, power supplies and power routing that is already in place. The TPC uses a water-cooled copper envelope on the front-end card to keep the temperature of the detector close to 21 °C. For the TPC the existing cooling can handle up to 35 mW/ch whereas the power supply and distribution for the MCH is limited to about 13 mW/ch.

2.3.3.6 Slow-Control

Assuming the default start-up configuration of the device is defined with settings that are close to what the TPC and MCH requires, there would not be a need to do many configuration changes after the device has cold booted. For this reason, a low-speed configuration is sufficient. The simplest option is to use I^2C (Inter-Integrated Circuit) as it provides multi-drop communication both ways using only two signal, which minimizes the amount of communication lines going to and from the SAMPA. In case of the MCH, this is beneficial as several devices are sharing a cable with a limited amount of communication lines. The GBT-SCA provides an I^2C master that can operate at 1 Mbps, which should be sufficiently fast enough. If faster communication is needed, another option is to include the outgoing slow-control data in the serial uplink path and add an extra serial input for reception of slow-control data. Since the differential serial-downlink can also be multidrop connected to several devices, it does not require more communication lines than the I^2C solution.

2.3.3.7 Reliability

Since the electronics will be located in a radiation area with limited access, there is a need for high reliability of the equipment to avoid downtime and loss of experimental data. Special care needs to be taken to protect the system against potential corruption of control and data registers caused by radiation-induced Single Event Effects (SEE). This also includes self-recovery in case of unforeseen errors and the ability to turn off broken channels to avoid that they transmit garbage data.

2.3.3.8 Design for Testability

The total number of chips required to cover all the channels of the TPC and MCH is about 51 000. Accounting for about 15% in spares and 30% in loss due to yield problems, this totals about 83 000 chips. As multiple chips will be mounted on the same front-end board, the yield of the front-end boards will be low if significant amounts of chips with manufacturing defects are mounted. For the TPC, with five chips on the front-end board, the probability of having a board with one or more chips with manufacturing defects is 41% if the probability of a chip having a manufacturing defect is 10%. The high number of chips that need to be tested requires an automatic testing procedure. The combined test time for a PASA and ALTRO chip for the original commissioning of the TPC was about 2 min, which if applied to the SAMPA chip would mean 3.8 months of continuous testing without parallelization. Effective ways of testing the device is therefore necessary. A possible option is to do additional testing at the manufacturing plant on the wafer level with automated test equipment to filter out bad devices before they are packaged. This comes at an additional cost, though it saves some time on the final testing, as the number of devices that needs to be tested at the packaged level will be fewer. Additionally there is also some cost savings as the number of devices that needs to be packaged is lower.

2.3.4 *TPC Specific Requirements*

The main TPC digital requirements were altered during the final submission phase of the second prototype due to new simulation results. As there are no major digital feature changes between the second prototype and the final device and since most of the digital implementation choices are based on the original requirements, the original requirements will also be presented here for clarity.

2.3.4.1 Digital Filter

The signals from the GEM detector do not feature a tail like on the MWPC and so the necessity for a tail cancellation filter is no longer present.

The baseline correction of non-systematic effects might still be useful, this is where a moving average value is subtracted from the current value to remove perturbations of the baseline. This filter was present in the ALTRO, but could not be utilized due to sudden swings in the baseline, which caused the calculation to fail and so also the zero suppression.

2.3.4.2 Other Compression Scenarios

In detectors like the TPC, the data output contains primarily small sample values, while large values are more rare. The distribution of sample values is approximately exponential [21]. By employing Huffman coding [22] the data size can be reduced by assigning shorter codes to frequent values and longer ones for more rare values. For occupancies above 25%, the Huffman coding will yield better compression factor than zero suppression [18], at lower occupancies the zero suppression is more efficient, though the Huffman coding is lossless so signals that are close to the noise will be kept, improving later charge calculations.

2.3.4.3 Readout Ordering

The charge of a track hitting the pad plane might be spread over multiple pads and so the signals from several pads are needed to find the centre of the charge. As the pads where a charge is spread over might not all be pads that are connected to one SAMPA, the calculation needs to be done later where the information from several SAMPAs is available. The central position and the total charge of each cluster hitting the pad plane is therefore planned to be calculated in the CRU. By recording only the position and charge of a cluster, the amount of data that needs to be stored for analysis is greatly reduced. This would though require that the data from each channel connected to a pad row arrive together in the CRU. The more out of order the data arrives, the more buffering is needed to get the data aligned. The possibility to configure a predictable order in which pad arrives before another is beneficial. As the mapping of which pads are connected to which channel is not the same in all positions of the detector, there would be a need for changing the predefined ordering also.

2.3.4.4 Interfaces

The GBTx provides 10 serial uplinks at 320 Mbps when Front-End Card (FEC) is applied, or 14 links when operating in the Wide Bus Mode where it is disabled. With FEC, several errors in a packet can be recovered in case of corruption during transmission due to either noise or radiation. Using two GBTx chips provides a maximum uplink bandwidth of 6.4 Gbps or 8.96 Gbps without forward error correction. Limiting to two GBTx chips per board will be done due to budgetary reasons. The new front-end cards will replace the existing ones, which each connects to 160 front-end channels and each front-end card will generate 16 Gbps of data. To fit all the data in the allotted bandwidth, the compression with lossless encoding must be better than 2.5 and with zero suppression the occupancy must be lower than 40%.

2.3.5 TPC Specific Requirements, Revised Architecture

A well-known issue with GEM detectors is that they possess a common-mode effect, which results in additional noise contributions if proper corrective baseline restoration has not been done. The effect results from the capacitive coupling between the amplification structure and the readout pads. When an avalanche occurs in the GEM foils, the current introduces a voltage drop on the GEM electrodes leading to induction of a correlated common-mode signal with opposite polarity on all anode pads facing the GEM electrode [23].

The designed filtering structures present on the SAMPA were only found to partially restore the performance [23]. Better performance can be achieved if information from a larger number of pads can be correlated, this correction would though need to be done on the CRU with raw uncompressed data.

2.3.5.1 Direct Readout Serialization

The SAMPA has the possibility to send raw samples or lossless encoded samples in the packet-based mode, but to be able to perform the common-mode correction on the CRU the data from all channels need to be in parallel. As the SAMPA sends one packet after another, this would require the need for a large amount of buffering on the CRU. A CRU can be connected to up to 1600 channels, which if no compression is used would require 16 Mbit of buffer memory. With lossless compression the size could be reduced, but would instead require extra logic for the decoding.

At the submission of the second prototype of the SAMPA a test mode was available that could serialize all the data from the 32 channels when running with a serial speed that is 32 times the sampling speed. The data is output on 10 serial links where the 10-bit samples from each channel are sent consecutively in one sampling period.

Running the SAMPA at the originally envisioned sampling speed of 10 MHz would though produce 16 Gbps of raw data from the five devices, whereas the two GBTx devices can only forward 6.4 Gbps in the normal operation mode with FEC, or 8.96 Gbps in the Wide Bus Mode where the FEC is turned off [24]. It is thus necessary to halve the sampling frequency to 5 MHz to avoid doubling the number of GBTx devices and optical links. An uplink speed of only 8 Gbps is then required. Simulations have proved that operating at 5 MHz does not degrade the detector performance [23].

As one of the SAMPAs will be connected to two GBTx, which might connect to different CRUs, there also needs to be an option to serialize half the channels on five links.

2.3.6 *MCH Specific Requirements*

With a drift length of only 2.5 mm in the detector, there are no issues with pileup from multiple events for the MCH like there is for the TPC. It also means the signal from an event will only be a single pulse of a few samples for the pads that are hit. This means the data amount per device will be very small, which requires another readout architecture than the TPC to keep the amount of GBTx devices and CRUs low.

2.3.6.1 Daisy Chaining

The MCH hit rate should be low enough so the data traffic from one chip can be routed through only one serial output link with a bandwidth of 40 Mbps. One of the two chips on the front-end card sends its data-readout stream to the other chip on the same front-end card, which then merges the data stream of its neighbour to its own data stream. In this way, each front-end card has only one single serial output link of 80 Mbps. This effectively halves the number of GBTx and optical links needed, significantly reducing the cost of the installation.

2.3.6.2 Data Compression Using Zero Suppression and Cluster Summing

Two devices of 32 channels each operating at 10 MHz produce in total 6.4 Gb/s of raw data. To reduce the data amount below the serial output link bandwidth of 80 Mb/s an efficient compression must be employed. Due to the bandwidth limit, conventional zero suppression might not be sufficient. As there is little probability of pileup in the MCH detector, it is enough to know the length of the pulse and the total charge to be able to reconstruct the pulse. Sending only the sum and length of the sample values representing the pulse, together with the time of arrival will significantly reduce the bandwidth needed. Further analysis is done at the end of this chapter.

2.3.6.3 Physical Size

The existing detector implementation will not be modified and such, the physical layout of the front-end boards, the location, and the connections to the chambers will be kept unchanged. Two versions of the front-end boards exist for the previous implementation; MANU 12 and MANU 345. They have the same functionality, but differ in the form-factor due to differences in the geometrical constraints for the two most forward and three most backwards chambers (1,2 and 3,4,5 respectively). The MANU 12 boards have a size of 23 mm × 63 mm and the MANU 345 have a size of 32 mm × 50 mm, which sets a limit for the size of the chip to about 15 mm × 15 mm. This leaves some room for routing of traces on the short sides for the MANU 12 boards and leaves enough space for placing the voltage regulators.

For the digital design, the size constraint mainly limits the amount of memory available for buffering of events and the number of power, input and output pins that can be positioned around the edges of the silicon die.

2.3.7 *Other Detectors*

As the project evolved, other detectors have shown an interest in using the device for their detector. The primary requirements of the device is set by the requirements of the TPC and MCH, though the development has tried to encompass the needs of the other groups as far as feasible without compromising the original requirements.

2.3.7.1 RHIC STAR

The Solenoid Tracker at RHIC (STAR) detector is one of four experiments at the Relativistic Heavy-Ion Collider (RHIC), located at Brookhaven National Laboratory in Brookhaven, USA. The primary physics goal for STAR is to study the formation and characteristics of QGP, similar to the ALICE detector at the LHC. The STAR detector uses a TPC as the main tracking detector. [27] The readout system of the TPC is based on MWPCs with readout pads. The readout pads are divided into 12 circular sectors on each side, where each sector is divided into an inner and outer sector. The pads in the inner sector do not provide uniform coverage at all radii like the outer sectors, as there is a spacing of about 5 times the pad height between rows,

Table 2.4 MPD [25, 26] and STAR [27, 28] front-end chip requirements

Specification	MPD	STAR iTPC
Number of pads	95 232	84 000
Polarity	Positive	Positive
Maximum event rate	<7 kHz	<3 kHz
Pad occupancy	<10%	~14%
Noise (ENC)	<1000 *e*	<700 *e*
Signal-to-noise ratio	30:1	20:1
Peaking time (t_s)	~180 ns	~180 ns
Dynamic range	10 bits	10 bits
Sampling rate	10 MSps	~10MSps
Power consumption (per ch)	<100 mW	<55 mW
Detector capacitance (C_d)	10 pF–20 pF	~15pF
Detector gas amplification factor	$\sim 10^4$	3770 inner
Drift time	<30 µs	~40 µs

causing loss of track information. The STAR TPC plans to upgrade the inner sector to twice the number of pads [28]. As can be seen from Table 2.4 the specifications for the upgraded system are close to the ALICE TPC requirements listed in Table 2.3.

2.3.7.2 NICA MPD

The Nuclotron-based Ion Collider fAcility (NICA) project is under construction at the Joint Institute for Nuclear Research (JINR) in Dubna, Russia [29]. The goal of the project is to study hot and dense baryonic matter in heavy-ion collisions, and to study spin structures in the collisions of polarized protons and deuterons. The collider can provide collisions of ions with atomic masses up to gold with energy range $\sqrt{s_{NN}} = 4 - 11$ GeV for gold, as well as polarized protons and deuterons with energy up to $\sqrt{s} = 27$ GeV for protons. One of three detectors at NICA is the Multi Purpose Detector (MPD), which is a detector specifically suited for heavy-ion collisions at high luminosities, similar to the ALICE detector. The main tracking detector in the MPD is a TPC based on MWPCs with cathode pad readout [25, 26]. The TPC is currently using the PASA and ALTRO devices and so a retention of the existing filtering functionalities of the ALTRO, particularly time based correction, pedestal correction and baseline correction is of interest. If the specifications for the MPD in Table 2.4 are compared to the ALICE TPC specifications in Table 2.3, we see that the MPD specifications are easier to fulfil than the ALICE ones. The noise requirement is lower, the power consumption requirement is lower, and the rest are comparable to the ALICE TPC.

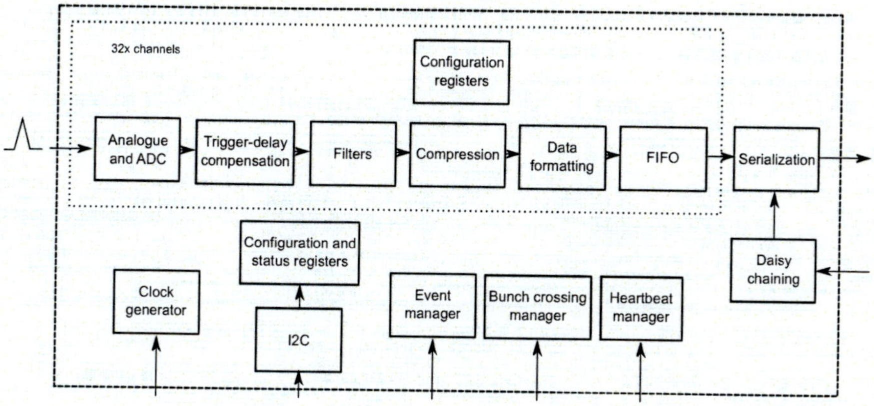

Fig. 2.5 Simplified block diagram of SAMPA architectural design

As both MPD and STAR are using MWPC TPCs, it is beneficial for them to retain the filtering and event handling features of the ALTRO that are more geared towards trigger based MWPCs. These include the tail cancellation filter, the baseline subtraction of systematic effects and the pre- and post-trigger handling.

2.4 Digital Specification

The SAMPA architectural design, as shown in Fig. 2.5, stays close to the architectural design of the ALTRO/S-ALTRO as was shown in Fig. 2.2. It contains a chain of signal conditioning filters, compression, data formatting and buffering per channel. Due to the operational requirement to run in continuous mode, the event-based buffer in the previous devices is replaced by a First-In, First-Out (FIFO) memory, able to hold an arbitrary amount of events, compared to the 4 events that was possible in the previous devices. The parallel-bus based interface used for readout and configuration is replaced by a serial interface for data readout and with I^2C for configuration. New methods of event management features are present that maintains better synchronization across the detector and conforms to the new trigger and acquisition requirements for the upgrade. An option to daisy chain several devices is available for detectors with low data amounts.

2.4.1 Event Management

The acquisition is based on triggering of events, where each trigger will start the acquisition of a programmable number of samples. The event control is handled globally in the device and all channels of the chip will start and stop acquisition at

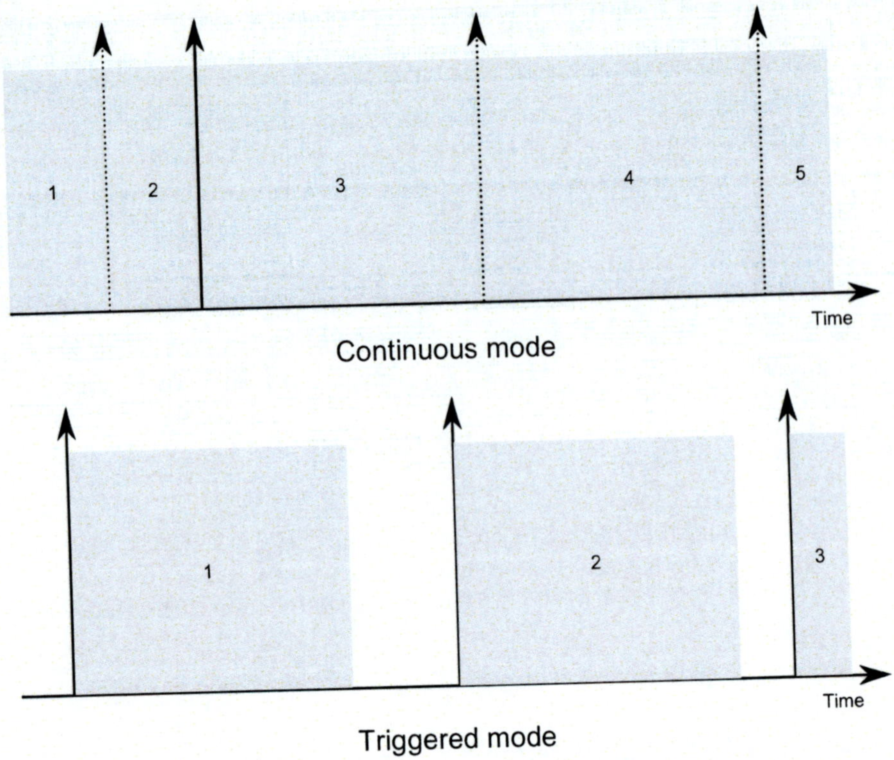

Fig. 2.6 SAMPA triggering modes. Gray areas indicate the acquisition time windows and are numbered. Solid arrows indicate external trigger, dotted arrow indicates automatic self-trigger

the same time. Processing and compression of the input samples does not affect the length of the event. The length of time that an acquisition lasts is only dependent on the configured sampling rate of the ADC and the programmable event length.

Each channel works on the incoming samples as a stream of data. It continuously processes the data through its filters, compresses it and writes it to a buffer memory. At the end of the event a header is generated that has information on the chip number, channel number, length of compressed data, event number etc. When the header has been generated and written to memory, then the event is ready to be sent out through the serial links.

The device supports two triggering modes; continuous mode and externally triggered. In continuous mode, a new event is started automatically when the previous one has ended. In triggered mode, an event is only started upon reception of an external trigger signal. At the end of an externally triggered event, the device goes back to being idle again until the next trigger arrives.

If an external trigger-signal is received in continuous mode, the current acquisition is ended and a new one is started immediately without loss of information. In this way, it is possible to align the events across different devices. An overview of the different triggering modes is shown in Fig. 2.6.

When using an externally provided trigger signal, it is often beneficial to be able to compensate for the delay between the trigger generator and the reception of the trigger signal on the device. This compensation is done by delaying the incoming data by the same amount as the trigger signal is delayed. This can also be used for looking at data prior to the triggering event by adding more delay. Implementing a programmable length delay chain of 192 samples satisfies the requirement of the TPC. The maximum delay configuration corresponds to a compensation of 19.2 µs at a sampling frequency of 10 MHz.

The particles travelling around the acceleration ring of the LHC are grouped in bunches and spaced apart by about 25 ns providing the 40 MHz that is distributed and used as the base frequency in many of the devices operating in the LHC, among other the GBTx. The SAMPA device has an internal 20-bit counter that runs at 40 MHz independently of the configured ADC sampling frequency. At the start of each new event, the value of this counter is captured and it is added to the header of that event as the event's bunch crossing id. If the ADC sampling clock is derived from the same source clock as the 40 MHz clock it is possible to use this value as an event number id by calculating :

$$event\,num = \frac{BXID}{event\,length \cdot \frac{40\,MHz}{F_s}} \tag{2.1}$$

Where $BXID$ is the bunch-crossing id from the header, $event\,length$ is the pre-programmed length of the event and F_s is the ADC sampling frequency.

On reception of an external bunch-crossing sync-trigger, the counter will be reset. This makes it possible to align the bunch-crossing id across different SAMPA chips.

For off-site synchronization, grouping of events, and to verify that a device is alive, a specially crafted heartbeat packet will be generated once a heartbeat trigger signal is received. The packet contains among other things the bunch-crossing id when the trigger arrived and the chip address.

2.4.2 Interfacing

The clock for the device is provided from the GBTx, which can provide clock frequencies relative to the link speed that is configured for a link, e.g. it will provide 80 MHz, 160 MHz and 320 MHz for 80 Mbps, 160 Mbps and 320 Mbps links. The operational clocks need to be derived inside the device. Since the clocks provided from the GBTx are synchronous to the LHC clock, the sampling clock of the ADC will then also be synchronous to this clock. For use with other detectors, an option for externally provided clocks is necessary.

The reset and trigger signal are connected to the GBTx and can be supplied to multiple devices in a multi-drop fashion. As the signals are pulse based, they can be transmitted at a lower data rate through the GBTx if needed. Higher speeds are only beneficial if there is need for better granularity in compensating for propagation delay of the optical links.

The number of active serial data links is configurable so that the detectors can choose the number of links needed to adapt to their needed bandwidth. Data from a neighbouring device can be routed through a device to further lower the number of used links. This requires an additional busy input and output signal to limit the data flow between the devices as a transmitting device will always have more memory for output buffering than a device that receives has for the input buffering.

2.4.3 Serial Link Data Protocol

As the GBTx forwards data without regard of the underlying protocol, there is no need to make the data protocol conform to any specific format. The data sent from the SAMPA over the serial links consists of a fixed length header and a variable length data payload, collectively referred to as a packet, see Fig. 2.7. Each packet corresponds to the event of one channel. The header is always 50 bits long, whereas the length of the payload for a given packet is given in the header of the packet as the number of 10-bit words. The format and specifications of the header field is given in Fig. 2.8 and Table 2.5. Each packet is identified by the type of packet (special or channel data, see Table 2.6), the device number of the originating device, the channel number of the originating channel and a bunch crossing identifier for when the event started. The device number could normally be deduced from the link number that the data originates from on the CRU, but not in the case of using daisy chaining as several devices can share the same link.

As the payload has a variable size, due to the compression employed, and since that size is given in the header, it is important to protect the header from errors to avoid that the receiving system loses synchronization when an error occurs in the payload size field. The header is thus protected with Single Error Correction, Double Error Detection (SECDED), which can correct for one error and detect two errors in the header. A parity bit for the payload data is also provided.

The channel buffers of the SAMPA might overflow if the collective output bandwidth is not greater than the amount of data produced. The channel buffers are split into a separate header and payload buffer so that if the payload buffer overflows it is possible to discard the payload for that packet and mark it in the header that the payload was discarded. In this way, the header can still be transmitted for the event to indicate to the receiver upstream that the data was lost.

The device has no busy output signal to indicate that its buffers are full, instead the CRU will monitor the received packet headers for indication of buffer overflows and forward the information to the Central Trigger Processor (CTP), which will throttle or stop the triggers.

Two special types of packages exists:

Heartbeat packet A specially crafted packet that contains a header with no payload, see Fig. 2.9. The header has its packet type marked as heartbeat, all the other

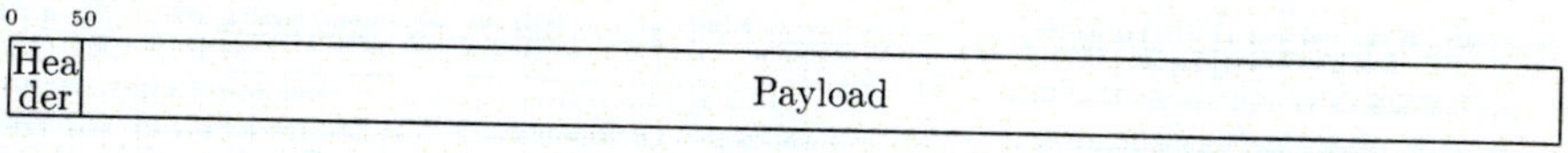

Fig. 2.7 Format of serial data. Length of payload is variable and encoded in the header information

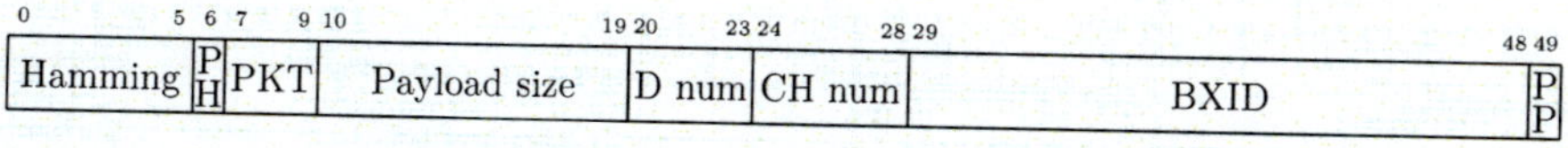

Fig. 2.8 Format of serial data header

Table 2.5 Protocol bit field descriptions for the serial data header

Name	Bits	Description
Hamming	6	Hamming parity bits
PH	1	Parity (even) of header including hamming
PKT	3	Packet type, see Table 2.6
Payload size	10	Number of 10 bit words in data payload, 11th bit is encoded in PKT
D num	4	Device number
CH num	5	Channel number
BXID	20	Bunch-crossing identifier (40 MHz counter)
PP	1	Parity (even) of data payload

Table 2.6 Packet type (PKT) encoding. Numbers in brackets are bit positions of the value from Table 2.5

Data	Heartbeat	Sync	Trigger during acq.	Payload discarded	Payload size [10]	PKT [2]	PKT [1]	PKT [0]
X					0	1	0	0
X					1	1	0	1
X			X		0	1	1	0
X			X		1	1	1	1
	X				0	0	0	0
X				X	0	0	0	1
		X			0	0	1	0
X			X	X	0	0	1	1

0–5	6	7–9	10–19	20–23	24–28	29–48	49
Hamming	PH	0x0	0x000	D num	0b10101	BXID	0

Fig. 2.9 Format of heartbeat packet

0–5	6	7–9	10–19	20–23	24–28	29–48	49
0x13	0	0x2	0x000	0xF	0x00	0xAAAAA	0

Fig. 2.10 Format of sync packet

fields are as normally except the channel id which is set to a fixed value of 21. The packet is initiated by a transmitting a heartbeat trigger signal to the device. At the moment when the trigger arrives, the bunch-crossing id is added to the packet and it is made ready to be transmitted. The heartbeat packets are only sent on the lowest number serial link and have higher priority than normal data packets, i.e. it will be sent after the transmission of the current packet has been completed on the link. When the device is daisy chained the packet will be passed through the chain. The packet are used for off-site synchronization during data analysis to group multiple events into a groups of events, additionally it is used by the data control system to verify that the device is alive.

Sync packet A specially crafted packet that contains a header with no payload, see Fig. 2.10. The header has its packet type marked as sync and the other fields have a fixed value that does not depend on the state of the chip or which chip it comes from. The Hamming code is still valid keeping the integrity of the header encoding intact. The sync packets are used by the receiving system to synchronize to the incoming data stream. When there is no data to send from the channel buffers or the daisy chain link, sync packets will be sent instead to make sure the receiving system keeps in sync with the data stream.

The 50-bit size of the header contributes very little to the overall bandwidth usage if zero suppression compression is in use. Assuming an event length of 1000 with 30% of data, the header will only contribute about 1.6%. On the other hand, the bandwidth overhead when using the MCH cluster summing compression is significantly higher as each pulse above threshold is compressed into only 4 words each. Keeping the header size in multiples of the data word size (10 bits) is beneficial for simplification of the data flow and so the minimum reduction of the header would be by 10 bits. This amount of reduction is not feasible without compromises. It would be possible to reduce the "number of words" field by 2-4 bits as the clusters are in multiples of 4 and as the number of clusters per event is expected to be small. A further reduction on the device number by 1-3 bits is possible, though another form of device number would need to be appended to each packet at the CRU to differentiate between packets coming from different links. In addition, the bunch-crossing id would need to be reduced by more than 3 bits, which with a reduction of 3 bits and an event length of 1024 would mean the counter would roll around every 32 packets. As discussed in Sect. 2.4.3.3 the bandwidth is though satisfactory for the MCH application.

2.4.3.1 Lossless Encoding

The Huffman coding uses variable length codewords [22]. Values that are present in the data with a higher frequency result in shorter codewords with fewer bits, whereas values with a lower frequency are coded in longer codewords. The Huffman coding creates codewords so that no codeword can be a prefix of another codeword, referred to as prefix codes. The codewords are uniquely identifiable in the stream and can be decoded without using any separators between the code boundaries. As each channel has a slightly different baseline, it is not efficient to make a table based on the raw input values from the ADC. However, since the signal from gas-based detectors primarily has gradual changes and few large changes, it is instead better to transmit the difference between a sample and the previous. The bulk of the values have then been found to lie between −10 and +10. The first value in an event needs to be stored unencoded so that it is possible to decode the following differentially encoded values.

With Huffman coding, the less probable value are represented by codes that have many more bits than the original 10 bits of the sample due to the requirement that one code cannot be the prefix of another. As the SAMPA device does not have the memory to store all codes, and since significant changes in occupancy can potentially increase the bandwidth usage too much if many long codes are suddenly more common, the SAMPA will instead use a limited programmable table of codes. The table will only have codes for the differentially encoded values that are close to zero, as these are generally the most common. For values that are outside of the table range the raw sample-value will be prepended by a special assigned codeword. When the stream is decoded this codeword is still unique, but it instructs the decoder to take the next 10 bits as a raw sample value and continue decoding from there. In a worst case scenario, where all values in an event is outside the table size, though this is not possible, the increase in size of a packet will only be about 40%, since the codeword used for values outside the table will generally be 3-4 bits long.

The bit-length of the payload for an event needs to end on a 10-bit boundary to conform to the rest of the readout architecture. As the header only dictates the number of 10-bit words in the payload and not the number of samples, the stream of data in the payload needs to end in a predictable manner for the Huffman decoder to decode the data properly. If the last sample in the event is always transmitted as if it was a value outside the table, then the decoder can have a rule that states that when a raw encoded sample ends within the last 10-bit word in the payload, then the rest of the bits should be discarded. This will only have a minor overall impact on the bandwidth.

2.4.3.2 Bandwidth TPC

Since tracks in the TPC are more plentiful at lower angles, there will be more data on the channels that are closer to the centre of the barrel than for the outer sections.

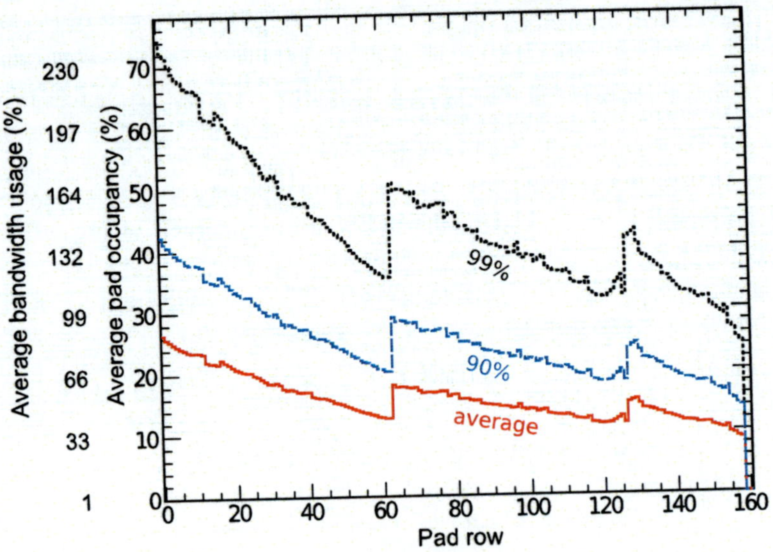

Fig. 2.11 Expected pad occupancies extrapolated from Run 1 data [18]. The red graph indicates the average pad occupancy on a given row for an average event. 90% of all events will have an average pad occupancy lower than the blue graph. 99% of all events will have an average pad occupancy lower than the black graph. Pad number goes from inner to outer radial row. On left side the corresponding expected bandwidth usage with 4 serial links at 320 Mbps is shown

The expected average equivalent pad occupancies[2] can vary from 9 to 26% from outer to inner row with an average of about 15%, as can be seen in Fig. 2.11. Some events may occasionally reach occupancies of up to 80%, far surpassing the available SAMPA bandwidth.

No simulated data for Run 3 is currently available, but from data taken in Run 1 the average cluster size when using zero suppression was found to be 6.5 samples at a sample rate of 10 MSPS. Together with the average occupancy, an estimate of the average bandwidth can be calculated.

$$BW = (H_b + (D_b \cdot P_{occ} \cdot N_{clusters})) \cdot f_{pkt} \cdot N_{ch} \tag{2.2}$$

BW_x is the bandwidth given in bps. H_b represents here the 50 bits of the header, D_b is the number of bits needed to represent a cluster, P_{occ} is the average occupancy, and N_{ch} is the number of channels on a chip. $N_{clusters}$ is the average number of clusters in an acquisition event, which is the configured length of an acquisition event divided by the average cluster size. f_{pkt} is the packet rate i.e. the sample rate divided by the acquisition event length.

[2] Pad occupancy is the fraction of samples within a given time window exceeding the zero suppression threshold. An equivalent pad occupancy is used, as there is an average pileup from five interactions in the detector for any given event.

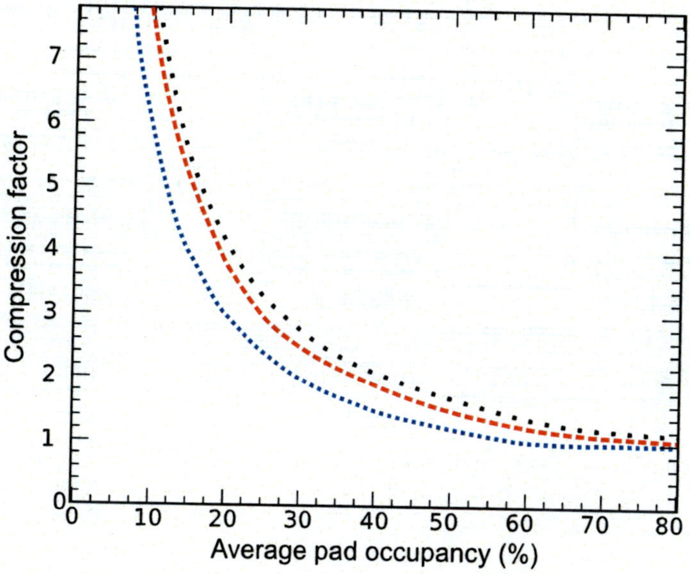

Fig. 2.12 Zero suppression compression versus occupancy for equidistant clusters of size 3 (blue), 6.5 (red) and 10 (black)

When using zero suppression with run-length encoding to compress the data between the clusters, a time stamp is needed to indicate the position for where the cluster starts in relation to the trigger. An additional word is also needed to record how many samples there are in the cluster. This is together encoded into two ten-bit words. With an average cluster size of 6.5 bits, the average amount of data per cluster is 85 bits. With a configured event length of 1023 samples, the average bandwidth usage for the innermost pad row is 1.1 Gbps, which is 86% of the available bandwidth when using four links at 320 Mbps. The maximum average occupancy that can be handled is about 30% due to the overhead of the zero-suppression compression.

Due to the higher rate of pile-up of clusters in Run 3, the average cluster length can be expected to be higher, which will overall improve the available bandwidth. Though increasing the cluster size from 6.5 to 10 only reduces the bandwidth by about 8%. The formula also overestimates somewhat at lower occupancies since the minimum size of a cluster for a shaping time of 160 ns and sampling frequency of 10 MHz is three samples. Figure 2.12 shows the compression factor versus occupancy for clusters of 3, 6.5, and 10. Compression needs to be higher than 2.5 to fit in the allotted bandwidth.

Since one in ten events will have an event size about twice of the average, a SystemC simulation framework was designed [30] to determine the amount of buffering needed on the SAMPA to handle these fluctuations. Results showed that by using a 4 kwords buffer, there would be a probability of $\sim 10 \times 10^{-16}$ of having a lost event due to buffer overflow in the 1000 innermost SAMPA devices [31]. The 6 kwords buffer implemented in the SAMPA was kept to avoid the additional risk and work of reducing the memory size.

2.4.3.3 Bandwidth MCH

The MCH is targeting an interaction rate of 100 kHz with 9% channel occupancy.[3] With a pulse shaping time of 300 ns, the average number of samples above the zero suppression threshold is 10 when the sampling frequency is 10 MHz. The bandwidth available per front-end card is 80 Mb/s, which is shared between the two chips and gives 40 Mb/s per chip.

As described earlier in Sect. 2.4.3, the SAMPA data transmission operates on a packet basis where a packet contains a header of 50 bits and a payload of compressed sample data. The amount of time-bins per packet is programmable up to 1023 bins. If no data is above the zero suppression threshold for the full duration of a packet, the chip would normally send only the header with an empty payload, but it is possible to set the chip to suppress these packets to save bandwidth.

The data rate for each of the possible operations modes of the MCH can be calculated directly, if we assume that there is no noise that creates false clusters.

$$BW_{trigger\ wo/empty} = H_b \cdot f_i \cdot N_{ch} \cdot P_{occ} + D_b \cdot f_i \cdot N_{ch} \cdot P_{occ} \tag{2.3}$$

$$BW_{trigger\ w/empty} = H_b \cdot f_i \cdot N_{ch} + D_b \cdot f_i \cdot N_{ch} \cdot P_{occ} \tag{2.4}$$

$$BW_{cont\ wo/empty} = H_b \cdot f_{pkt} \cdot N_{ch} \cdot (1 - (1 - P_{occ})^{\frac{f_i}{f_{pkt}}}) + D_b \cdot f_i \cdot N_{ch} \cdot P_{occ} \tag{2.5}$$

$$BW_{cont\ w/empty} = H_b \cdot f_{pkt} \cdot N_{ch} + D_b \cdot f_i \cdot N_{ch} \cdot P_{occ} \tag{2.6}$$

Where $BW_{trigger\ wo/empty}$ in (2.3) is the bandwidth needed for triggered mode with suppression of empty packets, $BW_{trigger\ w/empty}$ in (2.4) is the bandwidth needed for triggered mode without suppression of empty packets, $BW_{cont\ wo/empty}$ in (2.5) is the bandwidth needed for continuous mode with suppression of empty packets, $BW_{cont\ w/empty}$ in (2.6) is the bandwidth needed for continuous mode without suppression of empty packets. BW_x is given in bps. H_b is the 50 bits of the header, f_i is the interaction rate, f_s is the sampling rate, f_{pkt} is the packet rate which is the sample rate divided by the programmed length of a time window, N_{ch} is the number

[3] Channel occupancy is the fraction of channels that register a hit per interaction.

Table 2.7 Data rates per SAMPA for MCH. The numbers assume no detector noise

Mode	Cluster compression	Send empty packets	Size of cluster (D_b) [bits]	Average packet length [bits]	Data rate [Mb/s]	Max occupancy (%)
Triggered	None	No	120	15	49	7.4
Triggered	None	Yes	120	61	195	0
Triggered	Cluster sum	No	40	8	26	14
Triggered	Cluster sum	Yes	40	54	172	0
Continuous	None	No	120	138	44	8.1
Continuous	None	Yes	120	158	51	6.3
Continuous	Cluster sum	No	40	67	21	20
Continuous	Cluster sum	Yes	40	86	28	18.8

of channels, P_{occ} is the occupancy and D_b is the number of bits needed to represent a cluster.

When using zero suppression with run-length encoding to compress the data between the clusters, a time stamp is needed to indicate the position for where the cluster starts in relation to the trigger. An additional word is needed to record how many samples there are in the cluster. This is together encoded into two ten-bit words. In normal zero suppression mode the samples in the cluster themselves are not compressed so the size would be 10 bits for the time stamp plus 10 bits for the cluster size plus 10 bits × 10 bits for the cluster itself, for a total of 120 bits. With the cluster sum compression, the samples in the cluster are summed together into a 20-bit word, so the total is then 40 bits.

Table 2.7 present the data-rates per SAMPA for MCH. With the estimated occupancy-rates of 9%, the only suitable modes are triggered mode with cluster sum compression and suppression of empty packets, or continuous self-triggering mode with cluster sum compression. These estimates assume there is no detector noise.

A drawback of the triggered mode is that the bandwidth usage is more sensitive to noise than the continuous mode. Since there are only a few channels that have data per event, there is a large probability that any noise induced pulses would occur in a channel that would otherwise not have had data for that event. A header would thus also need to be sent, which would otherwise have been suppressed. If we assume that 10% of the channels will have a cluster of noise for each trigger, then the bandwidth can be calculated as

$$BW_{trigger\ wo/empty\ w/noise} = H_b \cdot f_i \cdot N_{ch} \cdot (P_{occ} + P_{noise} - (P_{occ} \cdot P_{noise})) + D_b \cdot f_i \cdot N_{ch} \cdot (P_{occ} + P_{noise}) \tag{2.7}$$

and the bandwidth is then increased to 53 Mbps, which is more than the available bandwidth. Overlap between signal and the noise can be ignored as it amounts to about

$$P_{occ} \cdot P_{noise} \cdot \frac{f_i}{f_s} = 0,009\% \tag{2.8}$$

assuming that the noise pulse is only one sample long.

Continuous mode is less affected by noise, as can be seen in (2.9), and the bandwidth can be calculated to be 38 Mbps, which is within the available bandwidth. The increase in bandwidth due to noise for continuous mode is 81% compared to the triggered mode, where the increase is 100%.

$$\begin{aligned} BW_{cont\ wo/empty\ w/noise} = {} & H_b \cdot f_{pkt} \cdot N_{ch} \cdot \\ & (1 - (1 - (P_{occ} + P_{noise} - (P_{occ} \cdot P_{noise})))^{\frac{f_i}{f_{pkt}}}) + \\ & D_b \cdot f_i \cdot N_{ch} \cdot (P_{occ} + P_{noise}) \end{aligned} \tag{2.9}$$

References

1. H. Soltveit, J. Stachel, P. Braun-Munzinger, L. Musa, H. Gustafsson, U. Bonnes, H. Oeschler, L. Osterman, and S. Lang, "The PreAmplifier ShAper for the ALICE TPC detector," *Nuclear Instruments and Methods in Physics Research Section A: Accelerators, Spectrometers, Detectors and Associated Equipment*, vol. 676, pp. 106–119, Jun 2012.
2. R. Bosch, A. J. de Parga, B. Mota, and L. Musa, "The ALTRO chip: A 16-channel A/D converter and digital processor for gas detectors," *IEEE Transactions on Nuclear Science*, vol. 50, no. 6, pp. 2460–2469, Dec 2003.
3. B. Mota, "Time-Domain Signal Processing Algorithms and their Implementation in the ALTRO chip for the ALICE TPC," Ph.D. dissertation, Ecole Polytechnique, Lausanne, Switzerland, 2003. [Online]. Available: http://cds.cern.ch/record/636665.
4. R. E. Bosch, "Study and design of the ALICE TPC front-end and readout electronics for the CERN LHC," Ph.D. dissertation, Universidad Politécnica de Valencia, Valencia, Spain, 2003.
5. G. Trampitsch, "Design and Characterization of an Analogue Amplifier for the Readout of Micro-Pattern Gaseous Detectors," Ph.D. dissertation, Graz University of Technology, Graz, Austria, 2007. [Online]. Available: http://cds.cern.ch/record/1472334.
6. P. Aspell, M. D. Gaspari, H. França, E. G. García, and L. Musa, "Super-Altro 16: A front-end system-on-chip for DSP based readout of gaseous detectors," *IEEE Transactions on Nuclear Science*, vol. 60, no. 2, pp. 1289–1295, Apr 2013.
7. M. De Gaspari, "Systems-on-Chip (SoC) for applications in High-Energy Physics," Ph.D. dissertation, Inst. Appl. Math., Heidelberg, Germany, 2012. [Online]. Available: http://cds.cern.ch/record/1607133.
8. E. J. G. García, "Novel Front-end Electronics for Time Projection Chamber Detectors," Ph.D. dissertation, Polytechnic University of Valencia, Valencia, Spain, 2012. [Online]. Available: http://cds.cern.ch/record/1607134.
9. H. M. F. Santos, "Highly Integrated Mixed-Mode Electronics for the readout of Time Projection Chambers," Ph.D. dissertation, Technical University of Lisbon, Lisbon, Portugal, 2013. [Online]. Available: http://cds.cern.ch/record/1563856.

10. T. Naaranoja, "Digital Signal Processing for Particle Detectors in Front-End Electronics," Master Thesis, University of Helsinki, Helsinki, Finland, 2014. [Online]. Available: http://hdl.handle.net/10138/135874.
11. P. Moreira, R. Ballabriga, S. Baron, S. Bonacini, O. Cobanoglu, F. Faccio, T. Fedorov, R. Francisco, P. Gui, P. Hartin, K. Kloukinas, X. Llopart, A. Marchioro, C. Paillard, N. Pinilla, K. Wyllie, and B. Yu, "The GBT project," in *2009 Topical Workshop on Electronics for Particle Physics Conference Record*, vol. CERN-2009-006, 2009, pp. 342–346. [Online]. Available: http://cds.cern.ch/record/1235836.
12. The ALICE TPC Collaboration, *ALICE TPC Readout Chip User Manual v. 0.2*, CERN, 2002. [Online]. Available: http://ep-ed-alice-tpc.web.cern.ch/ep-ed-alice-tpc/doc/ALTRO_CHIP/UserManual_draft_02.pdf.
13. J. Troska, "The Versatile Transceiver Proof of Concept," in *2009 Topical Workshop on Electronics for Particle Physics Conference Record*, vol. CERN-2009-006, 2009, pp. 347–351. [Online]. Available: https://cds.cern.ch/record/1235837.
14. J. Mitra, S. Khan, S. Mukherjee, and R. Paul, "Common Readout Unit (CRU) - A new readout architecture for the ALICE experiment," *Journal of Instrumentation*, vol. 11, no. 03, p. C03021, Mar 2016.
15. A. Caratelli, S. Bonacini, K. Kloukinas, A. Marchioro, P. Moreira, R. D. Oliveira, and C. Paillard, "The GBT-SCA, a radiation tolerant ASIC for detector control and monitoring applications in HEP experiments," *Journal of Instrumentation*, vol. 10, no. 03, p. C03034, Mar 2015.
16. H. D. H. Herrera, "Noise and PSRR improvement technique for TPC readout front-end in CMOS technology," Ph.D. dissertation, Escola Politécnica, São Paulo, Brazil, 2015. [Online]. Available: http://www.teses.usp.br/teses/disponiveis/3/3140/tde-05072016-151016/.
17. The ALICE Collaboration, "Upgrade of the ALICE read-out and trigger system," CERN, Tech. Design Rep. CERN-LHCC-2013-019, ALICE-TDR-015, 2013. [Online]. Available: http://cds.cern.ch/record/1603472.
18. The ALICE Collaboration, "Upgrade of the ALICE Time Projection Chamber," CERN, Tech. Design Rep. CERN-LHCC-2013-020, ALICE-TDR-016, 2013. [Online]. Available: http://cds.cern.ch/record/1622286.
19. G. Traversi, M. Manghisoni, V. Re, L. Gaioni, and L. Ratti, "Perspectives of 65 nm CMOS technologies for high performance front-end electronics," in *Vertex 2012*, 2012. [Online]. Available: http://cds.cern.ch/record/1693462.
20. M. Manghisoni, L. Gaioni, L. Ratti, V. Re, and G. Traversi, "Assessment of a Low-Power 65 nm CMOS Technology for Analog Front-End Design," *IEEE Transactions on Nuclear Science*, vol. 61, no. 1, pp. 553–560, Feb 2014.
21. C. Patauner, "Lossy and lossless data compression of data from high energy physics experiments," Ph.D. dissertation, Graz University of Technology, Graz, Austria, 2011. [Online]. Available: http://cds.cern.ch/record/1433839.
22. D. A. Huffman, "A method for the construction of minimum-redundancy codes," *Proceedings of the IRE*, vol. 40, no. 9, pp. 1098–1101, Sept 1952.
23. H. Appelshäuser, M. Arslandok, M. Ivanov, C. Lippmann, and J. Wiechula, "Readout scheme of the upgraded ALICE TPC," CERN, Tech. Rep. ALICE-PUBLIC-2016-006, Nov 2016. [Online]. Available: https://cds.cern.ch/record/2231785.
24. P. Moreira, S. Baron, S. Bonacini, O. Cobanoglu, F. Faccio, S. Feger, R. Francisco, P. Gui, J. Li, A. Marchioro, C. Paillard, D. Porret, and K. Wyllie, "The GBT-SerDes ASIC prototype," *Journal of Instrumentation*, vol. 5, no. 11, p. C11022, nov 2010.
25. A. V. Averyanov, A. G. Bajajin, V. F. Chepurnov, G. A. Cheremukhina, O. V. Fateev, A. M. Korotkova, F. V. Levchanovskiy, J. Lukstins, S. A. Movchan, S. V. Razin, A. A. Rybakov, S. V. Vereschagin, Y. V. Zanevsky, S. A. Zaporozhets, and V. N. Zruyev, "Readout system of TPC/MPD NICA project," *Physics of Atomic Nuclei*, vol. 78, no. 13, pp. 1556–1562, Dec 2015.
26. A. Averyanov, A. Bajajin, V. Chepurnov, V. Chepurnov, G. Cheremukhina, S. Chernenko, O. Fateev, Y. Kiriushin, A. Korotkova, F. Levchanovskiy, J. Lukstins, S. Movchan, A. Pilyar, S. Razin, A. Ribakov, V. Samsonov, S. Vereschagin, Y. Zanevsky, S. Zaporozhets, and V. Zruyev, "Time projection chamber for multi-purpose detector at NICA," *Acta Physica Polonica B Proceedings Supplement*, vol. 9, no. 2, p. 155, 2016.

27. M. Anderson, J. Berkovitz, W. Betts, R. Bossingham, F. Bieser, R. Brown, M. Burks, M. C. de la Barca Sánchez, D. Cebra, M. Cherney, J. Chrin, W. Edwards, V. Ghazikhanian, D. Greiner, M. Gilkes, D. Hardtke, G. Harper, E. Hjort, H. Huang, G. Igo, S. Jacobson, D. Keane, S. Klein, G. Koehler, L. Kotchenda, B. Lasiuk, A. Lebedev, J. Lin, M. Lisa, H. Matis, J. Nystrand, S. Panitkin, D. Reichold, F. Retiere, I. Sakrejda, K. Schweda, D. Shuman, R. Snellings, N. Stone, B. Stringfellow, J. Thomas, T. Trainor, S. Trentalange, R. Wells, C. Whitten, H. Wieman, E. Yamamoto, and W. Zhang, "The STAR time projection chamber: a unique tool for studying high multiplicity events at RHIC," *Nuclear Instruments and Methods in Physics Research Section A: Accelerators, Spectrometers, Detectors and Associated Equipment*, vol. 499, no. 2, pp. 659–678, 2003.
28. STAR Collaboration, "A Proposal for STAR Inner TPC Sector Upgrade (iTPC)," BNL, Tech. Rep. SN0619, 2015. [Online]. Available: https://drupal.star.bnl.gov/STAR/starnotes/public/sn0619.
29. V. Kekelidze, A. Kovalenko, R. Lednicky, V. Matveev, I. Meshkov, A. Sorin, and G. Trubnikov, "Status of the NICA project at JINR," *EPJ Web of Conferences*, vol. 138, p. 01027, 2017.
30. H. R. Olsen, "SystemC simulation of the future SAMPA ASIC for use in the ALICE experiment in run 3," Master Thesis, Bergen University College, Bergen, Norway, 2015.
31. C. Lippmann, "TPC readout for Run 3," Meeting on TPC specifications for SAMPA. [Online]. Available: https://indico.cern.ch/event/441670/contributions/1093710/attachments/1145950/1642952/TPCU_Rdo_150828.pdf.

Chapter 3
SAMPA Chip Implementation

This chapter gives a more in-depth description of the implementation of the design, with a focus on the digital implementation. For a more in-depth description of the analogue design, see [1].

3.1 Architectural Overview

The SAMPA chip consists of 32 identical channels, each containing a CSA, a shaper, and an ADC, which together convert the input charge from the detector to a semi-Gaussian shape, and digitizes it with 10-bit precision. A block diagram is shown in Fig. 3.1. The digital section is implemented as a single block receiving data from all of the 32 ADCs. It contains a filtering, buffer, and compression-processing pipeline for each of the input channel, in addition to providing common control and monitoring solutions. Eleven serialization blocks interface with the buffers that are present at the end of the channel filtering pipelines to transmit the data on to the receiving system. The device is implemented in a 130 nm Taiwan Semiconductor Manufacturing Company (TSMC) Complementary Metal Oxide Semiconductor (CMOS) process where the die measures 9.5 mm × 8.9 mm without cut lines. The die has 311 connectible pads and is mounted in a 15 mm × 15 mm 320-pin Ball Grid Array (BGA) package.

3.1.1 Analogue Front-End

The front-end block consists of a positive/negative polarity CSA, a pole-zero cancellation network, a high pass filter, two bridged-T second order low pass filters and a non-inverting stage [2], as shown in Fig. 3.2 [1].

A. Velure, *A Digital Signal Processor for Particle Detectors*, Springer Theses,
https://doi.org/10.1007/978-3-030-71559-5_3

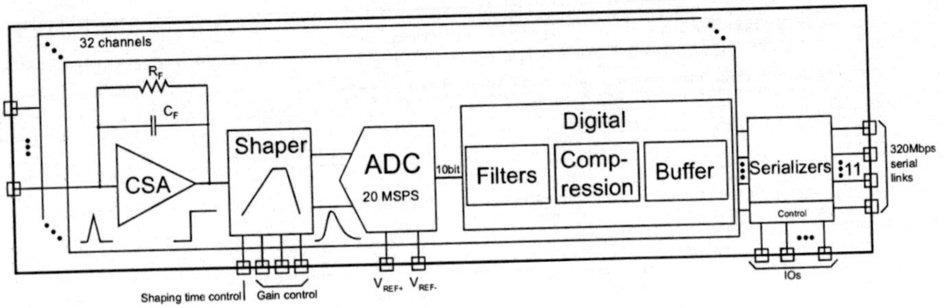

Fig. 3.1 Principal block diagram of the SAMPA

The charge-sensitive pre-amplifier integrates the input charge using a feedback capacitor (C_f). The parallel large-ohmic transistor (R_f) is used to discharge the capacitor, leading the signal back to the baseline level after the pulse with an exponential tail. Following the integration section is a pole-zero cancellation network to remove undershoots in the pulse shape.

After that follows a pulse shaping element, which reduces the spectrum of the signal resulting in a pulse with a 4th order semi-Gaussian shape. 4th order shaping is chosen because it offers a faster return to baseline as compared to lower order filters. This helps preventing pile-up effects when two or more pulses are closely spaced in time.

The second stage of the shaper is a fully differential second order bridged-T filter and includes a common-mode feedback network. The non-inverting stage adapts the DC voltage level of the shaper output to use the full dynamic range of the ADC. It consists of a parallel connection of two equally designed Miller compensated amplifiers.

Some of the parameters can be set externally using static control pins. The gain can be set to 20, 30 or 4 mV/fC, the shaping time can be set to either 160 or 300 ns, and the signal input polarity can be chosen.

Biasing of the analogue block is done from a common internal configurable reference voltage source (band-gap) that creates the required 450, 600 and 750 mV reference voltages. To be able to adjust the voltage reference, in case of large simulations versus production differences or to increase the yield, a 3-bit configuration is provided through a digital configuration registers that can trim each of the voltages ± 100 mV.

To reduce coupling of switching noise, the front-end has been separated into three independent power domains. The CSA and first shaper share one power domain, whereas the second shaper and the ADC have their own power domains. The separate ground and power nets are not merged inside, but brought to the outside of the device to individual pins such that it is up to the board designer to choose to externally combine the nets in the way that is most suitable to reduce noise coupling further if needed.

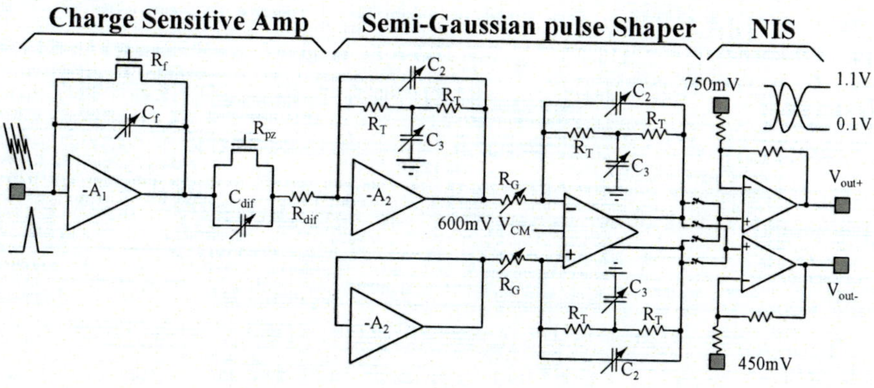

Fig. 3.2 Block diagram of the analogue front-end

3.1.2 Analogue to Digital Converter

The ADC is based on a split capacitor fully differential Successive Approximation Register (SAR) topology. The SAR ADC operates by doing several successive comparisons cycles to complete one conversion and so has a low operational speed, but requires a lower area compared to for instance pipelined ADCs like what was used in the previous ALTRO/S-ALTRO devices. This, in turn, gives lower overall power consumption. Recent reductions in the feature size of CMOS devices has made it possible to create SAR ADCs that can reach operational speeds and have an accuracy which is suitable for the SAMPA requirements [3]. The ADC for the SAMPA has a resolution of 10-bit and a sample rate of up to 20 MS/s. The block diagram of the ADC is shown in Fig. 3.3. The main parts of the circuit are the sample and hold input circuitry, the Digital to Analogue Converter (DAC) in the form of a binary weighted capacitive array, the comparator that compares the sample and hold value against the current value of the DAC, and the SAR control logic.

The SAR operation is based on a binary search algorithm and requires 12 cycles to complete a conversion. In the first cycle, the registers are reset, in the second the Most Significant Bit (MSB) is set to one. The register value is applied to the DAC and the sampled value is compared to the DAC value in the comparator. If the sampled value is lower than the DAQ value, the MSB is set low and the register is shifted right.

The digital design provides two clocks for each of the ADCs, a sampling clock and a clock for the SAR state machine. The task of the sampling clock is to restart the SAR state machines and make it ready for the next capture phase. The SAR state machine operates with a clock that is more than 16 times the sampling clock. An internal clock doubler for the SAR state machine clock makes sure the ADC can operate at 10 MS/s with an input clock of 80 MHz, suitable for the MCH.

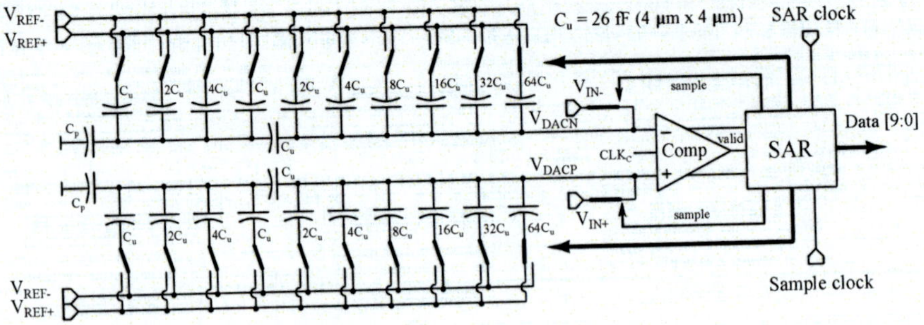

Fig. 3.3 Block diagram of the ADC

3.2 Digital Implementation

The task of the digital section is to administer the acquisition, filter the data that is acquired, compress it, and serialize it for transmission to an external device. A block diagram of the design is shown in Fig. 3.4. The filtering and compression chain for each channel is shown in the bottom part, the right part contains the serialization, the top centre right contains the daisy chaining option while the top centre and top left contains the event management and other elements that are common to the design. The task of the clock manager is to divide the input clocks down to the required internal frequencies and distribute them according to the setting configured on the external clock configuration pins. The clock manager provides both the required clocks for the digital section as well as for the ADCs. A reset manager is provided, which takes care of synchronizing the incoming reset signal to the appropriate clock domains. The slow control interface for the design is provided through an I^2C slave. It interfaces directly to a register bank, global to the design, which contains all the common configuration settings needed to operate the device.

The channel pipeline consists of a sequence of modules that filter and compress the data. The pre-trigger module is a small circular buffer that can be used to extend the pipeline and thus also delay the data in relation to the acquisition trigger, to compensate for any delay from the triggering source. The Baseline Correction I corrects for slow perturbations of the baseline of the signal. The Tail Cancellation can, as the name implies, cancel long tails of the signal, correct for peaking time of the pulse or act as a general filter. The Baseline Correction 2 is configured as a moving average filter and can remove faster variations in the baseline. The third Baseline Correction has the same function as the second, but uses a slope-based algorithm for correction of the baseline, which has inherently better stability. The baseline filtering is necessary for an efficient zero suppression and thus a high compression factor. The zero suppression indicates to the Data Formatting unit which samples are below a given threshold and therefore can be suppressed when using run-length encoding to compress the data. The Huffman module provides an alternative to the run-length encoding compression and gives a lossless compression of the data. The

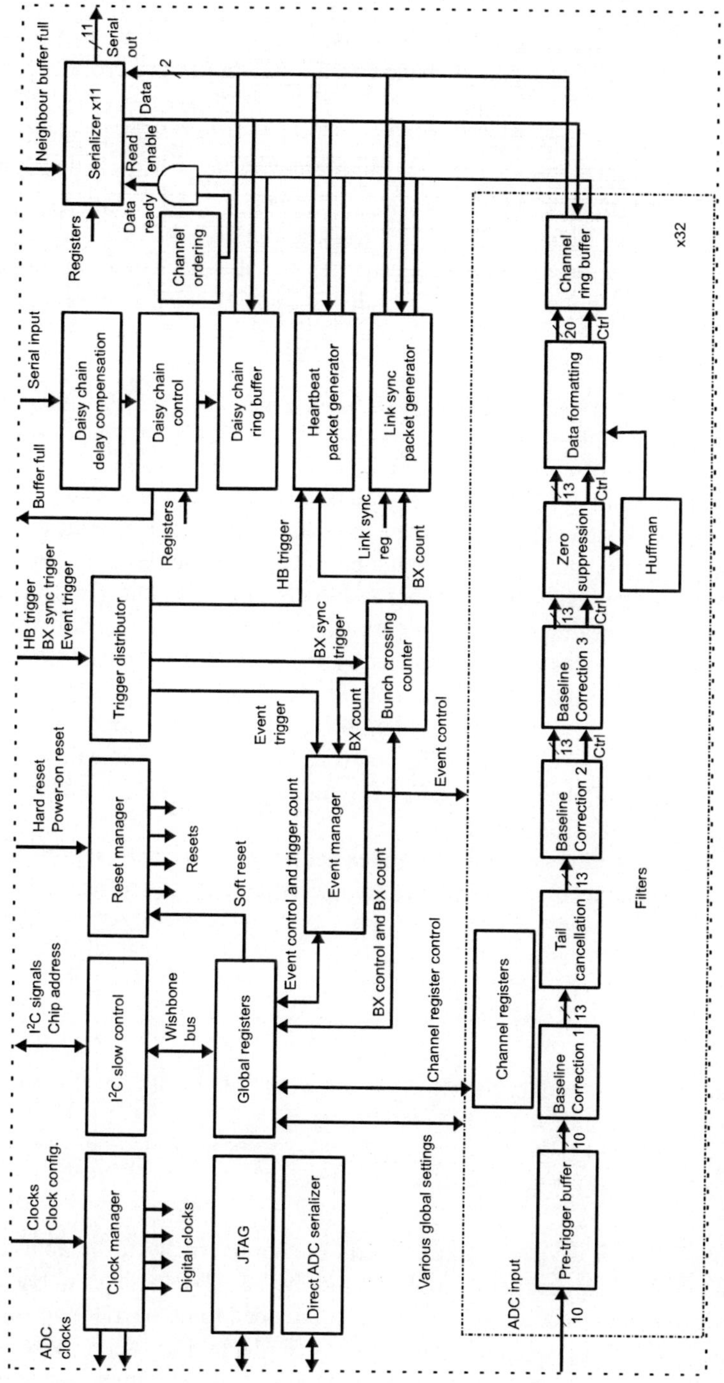

Fig. 3.4 Simplified block diagram of the digital design. Some independently operating test features are not shown for clarity

Data Formatting unit does the actual compression and assembling of the data into either run-length encoded data, with or without cluster sum compression, or Huffman data. The channel ring buffer consists of an input part and an output part. The input stores the compressed data from the Data Formatting unit in a ring buffer and attaches a header to the data once the event is completed. The output retrieves data from the ring buffer on request from the serialization module. Each channel has a register bank containing the configuration settings for its filters. The channel register bank is accessible through the global register.

The trigger distributor takes care of synchronizing the external trigger signals for each receiving domain. The event trigger signal is delivered to the event manager, which controls the triggering and distribution of event control signals to all the channels. An internal 20-bit counter, called the bunch crossing counter, is used for timestamping of events and can be reset from another external trigger signal (BX sync trigger). Heartbeat packets are used by the offline system to group events and check that the front-end device is alive. The transmission of a heartbeat packet is triggered by yet another external trigger signal (HB trigger).

Up to eleven serialization units can be enabled. Their task is to fetch packets from either the heartbeat generator, the daisy-chained link, either of the channels, or from the sync packet generator if no other packet is available. The serializers fetch packets from the channel in a round-robin fashion according to the channel number, but the order can be rearranged by configuring a new order in the channel ordering registers.

Several devices can be connected in series in a daisy-chained fashion. A separate delay compensated serial data input is available for this purpose, together with a flow control signal to limit the data flow. A daisy chain control module takes care of synchronizing to the incoming data and controlling the data flow, a separate daisy chain ring buffer takes care of storing the data in the buffer, making it ready to be retransmitted by the serializer.

For applications with less complex needs, the data can be serialized directly from the ADC inputs by an independently operating Direct ADC serializer module.

3.2.1 *System Services*

This section describes the clocking and reset functionality of the device.

3.2.1.1 Clock Management

The SAMPA device requires several clocks internally to be able to serialize the relatively slow parallel input data from the 32 ADCs at high output data rates. As mentioned in Sect. 2.4.2, the device is intended to be operated in conjunction with the GBTx chip, which has three options for data serialization speeds, 320, 160, and 80 Mbps. Clocks are supplied from the GBTx at the corresponding frequencies as well.

The ADCs require sampling clocks of up to 20 MHz and conversion clocks at 16 times the sampling clock speed. The sampling clock is used to start a new conversion of the analogue input signal into a digital signal, whereas the conversion clock is used for advancing the state machine that completes the conversion. Internally to the ADC the conversion clock is doubled, which makes it possible to operate with a sampling speed of 10 MHz and an input clock of only 80 MHz as required by the MCH.

For lowering power consumption and potentially generated noise from the digital, the processing of the sample data in the digital part happens at the selected ADC sampling speed as well. However, the sampling clocks supplied to the ADCs are from a different clock root than the one used for the digital. This is done to supply a cleaner clock to the ADCs in terms of skew and jitter. It also presents the opportunity to delay the sampling clocks for the ADCs so that the conversion does not occur at the same time as the digital part is active processing data.

The LHC operates on a synchronized clock, slightly higher than 40 MHZ, called the bunch-crossing clock. It is named so as the period is equal to the nominal time between the passings of two bunches of particles in the main accelerator ring. To synchronize the colliding of two bunches with the produced data in the detector, the SAMPA keeps an internal counter running at the bunch-crossing clock speed, which is used to mark when the first sample in a packet occurred. The clocks received from the GBTx are multiples of the bunch-crossing clock frequency and so an internal clock is derived from the input clock to operate as the internal bunch-crossing clock. Since the clock is a fixed frequency for all applications, in contrast to the ADC sampling clock, the bunch-crossing clock is also used for operation of the I^2C. This enables the I^2C to operate reliably at the highest speed supported by the GBT-SCA of 1 MHz.

In the section of the digital design that handles readout from the buffer memories and serialization of the data to the serial links, there is a need for a clock that is a fixed fraction of the serial clock. This fraction must also be a factor of 10 since the data is stored in 10-bit words, i.e. the clock needs to be either 1/10, 1/5, 1/2 or 1/1 of the serial clock speed. Since the buffer memories have a maximum operating speed lower than the maximum serializing speed of 320 MHz, the full speed cannot be chosen. Instead 1/2 of the serial speed is used due to the ease of generating a 1/2 clock from the input clock by halving, compared to using a 1/10 or a 1/5 clock, which will require a counter based division, which is more prone to SEE as well. A lower clock speed has the benefit of generating lower noise and power, but the circuitry operating on this clock domain is of a small size and physically located furthest away from the analogue section, so the influence is not significant.

In the section that handles writing to the buffer memory from each processing channel, a higher clock speed is needed as well because in some cases there is a need to store some control data at the same time as a sample is received. The same clock is used in this section as for the output section, since with the output clock as 1/2 the serializing clock it is always higher than two times the sample clock. This avoids a separate clock domain, as would be the case if 1/10 or 1/5 would be selected.

The main input serializing clock is primarily used only in the final serialization before the output links, where it serializes two bits at a time. Double-edged clocking or a Double Data Rate (DDR) cell would have been useful for the serialization and de-serialization parts of the design so that there would be one less clock domain to be distributed across the design. However, double-edged clocking, by using a rising and falling edge flip-flop together with a multiplexer, adds some additional issues when designing the scan chain capability,[1] as the scan-chain only operates with single-edged clocking. Those parts of the design would need to be excluded from the chain or some additional care would have to be taken to change the negative-edged flip-flops to positive-edged when the scan-chain functionality is enabled. When double-edged clocking is in use, there is also the possibility that the duty cycle would be asymmetrical for the output data, which could lead to capturing issues on the GBTx. The standard cell libraries for the process did not include DDR cells and a custom one was not designed due to lack of resources, though in hindsight this functionality could probably have been included in the custom SLVS driver that was created for the design.

The clock management is implemented as shown in Fig. 3.5 and is controlled by seven pins externally to the device. The design has been architectured in such a way that it is possible to always run with a fixed 40 MHz bunch-crossing clock and keep a sampling clock speed of either 5, 10 or 20 MHz independent of the input serial link clock being 80, 160 or 320 MHz. Both the sampling clock and bunch-crossing clock can be configured to be supplied from an external source. The sampling clock can additionally be configured to be derived from the bunch-crossing clock input if it is external or they can both be derived from the serial link clock input. A lookup table is used to convert the status of three of the control pins into the multiplexer configuration for the eight valid combinations of serial link speed and sampling speed as shown in the bottom left of Fig. 3.5.

One of the clock configuration pins is used as a main control for disabling the clock to a majority of the digital circuitry by clock gating. This is utilized when operating in the direct readout serialization mode[2] where the functionality of the design is reduced to the bare minimum to serialize data directly from the ADC. This saves power and reduces potentially generated noise.

All the required internal frequencies in the device are generated with ripple counters [4], i.e. the frequency is halved in subsequent steps to get the required frequency. For applications where there is a need for other frequencies, the sampling clock can be supplied externally instead. Since there were no requirements for specific sampling frequencies that could not be derived from the main clock by factors of two, a Phase Locked Loop (PLL) is not included due to the extra design complexity that this would involve.

[1] See Sect. 3.3.1 for more information about the scan chain implementation.

[2] See Sect. 3.2.6.3 for more information about the direct readout serialization mode.

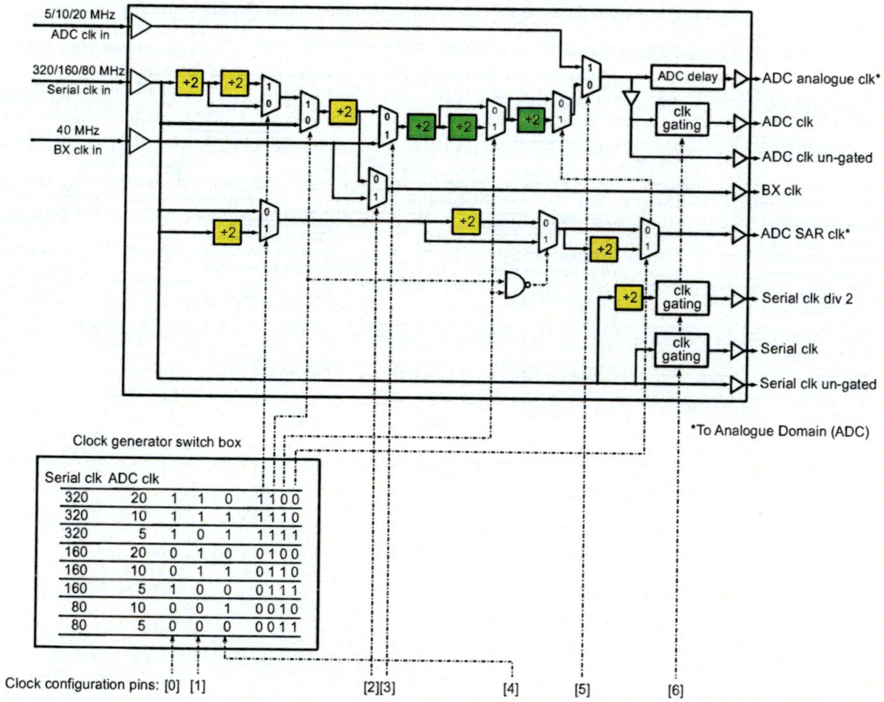

Fig. 3.5 Block diagram of the clock generation tree. All flip-flops have asynchronous clear. The flip-flops marked in yellow uses the primary reset signal from Fig. 3.7, while the green uses the ADC reset signal from the same figure

3.2.1.2 Reset Management

The primary purpose of the reset is to bring the device into a known state so that the device will start operating predictably. A reset signal can either be supplied externally from the device or it could be generated internally, this be either from a state machine that self-resets or from an analogue power-on-reset block that provides a reset signal once the power supply is found to be good. This design uses both methods. An input pin is dedicated to a hard-reset link, which can be controlled by a remote device. Additionally, there is a power-on-reset block present in the analogue section. The signal from the power-on-reset generator is brought out on a pin, which can be brought into the digital design via another pin if the user chooses to use the built-in power-on-reset functionality.

The design uses synchronous reset, but has an asynchronous assert, synchronous de-assert reset synchronizer on the input to avoid metastability problems and guarantee a long enough reset pulse. A benefit of using synchronous reset in the design is that the circuit will be completely synchronous and all the timing can be taken care of by the static timing analysis tool [5]. Synchronous reset also ensures that a reset can only happen on a clock edge and so the design is less prone to glitches on the

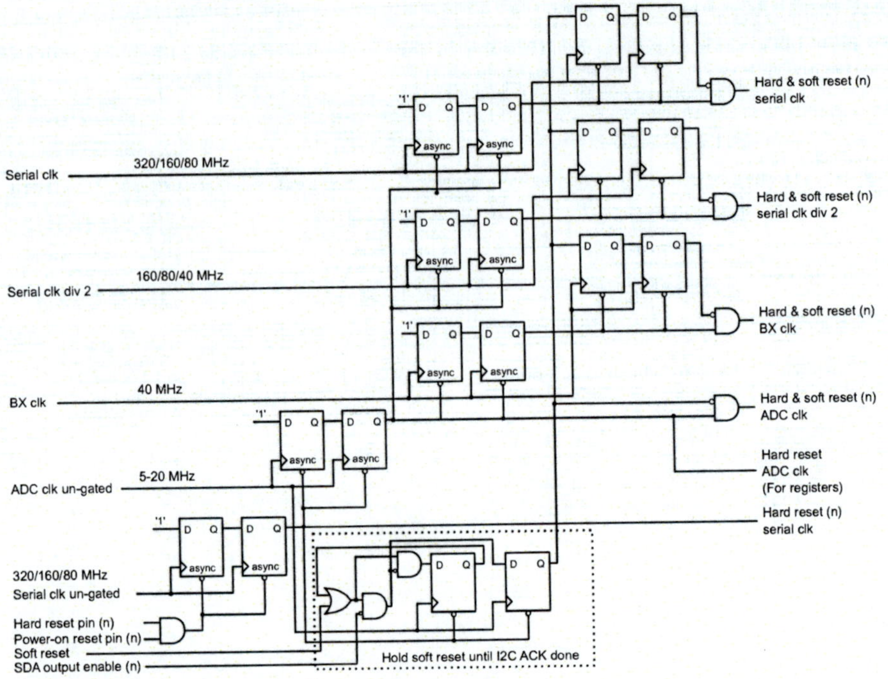

Fig. 3.6 Schematic diagram of the reset tree. (n) indicates active low signals. The registers are TMR protected, as further described in Sect. 3.4.1

reset line either from external causes or from radiation. An asynchronous reset has additionally the disadvantage that if the reset is de-asserted too close to the clock edge, the flip-flop might go metastable. It is difficult to guarantee that this will not happen when there could be different clocks in the design with different internal delays in relation to the reset signal. A drawback of synchronous reset is that the complexity of the reset tree increases as a separate reset is needed per clock domain and the release of the reset might need to be sequenced when bringing up the different clock domains. Additionally a clock always needs to be present for a reset to happen so care must be taken for gated parts of the design.

The reset tree is implemented as shown in Figs. 3.6 and 3.7. The input from the power-on-reset and the hard reset input is ANDed together directly to provide a common signal. By using asynchronous flip-flops in the first synchronizers, a clock does not need to be present at the moment of reset assertion. If some of the clocks used in the design is configured to be generated internally by the clock generator they will start up in order from fastest to slowest. As the faster domain might be ready before the first clock cycle in the slow domain, i.e. before the slow domain registers are reset, the values from the uninitialized register in the slow domain might be propagated to the fast domain which might bring the system into an unknown state. To avoid this, the fast domains are kept in reset until the slow domain is out of reset.

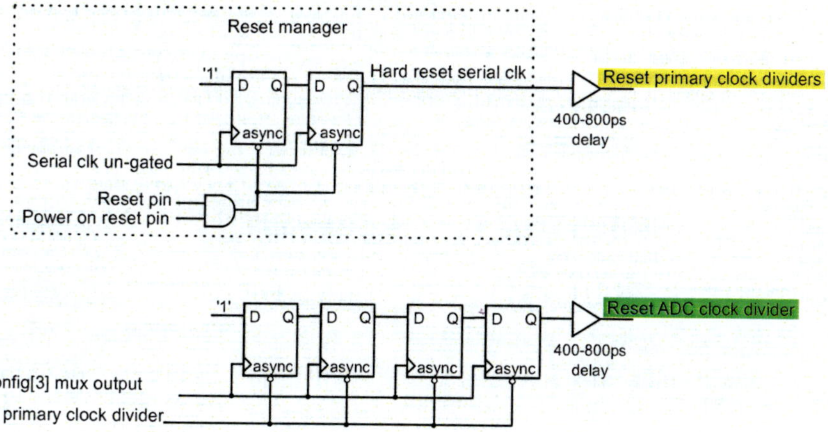

Fig. 3.7 Schematic diagram of the reset for the clock generation tree

A soft reset signal is provided through a global register accessible via the slow control. The soft reset will reset everything except thc clock generators and the configuration registers. In this way, the device can be reset to a known state without having to reconfigure it. The memories in the device are not reset or initialized in any way and will therefore retain their data even through a hard reset. The flip-flop circuitry connected to the soft reset line in Fig. 3.6 was added to make sure the I^2C acknowledge of the soft reset command is transmitted before a reset takes place.

When relying on the power-on-reset to get the device into a known state, the devices across a detector will likely not start up at exactly the same time as this depends on when the devices receive power, which can be difficult to synchronize if there are several power supplies that might have different loads. When deriving the sampling clock from a higher speed clock internally to the device, the phase of the sampling clock will be dependent on when the device was reset. If the devices have not started up at the same time, there will likely be a phase difference in the sampling clock between devices as well, which normally is undesirable. Since the reset manager has been made with a predictable constant start-up delay, the devices across a detector can be reset, and the sampling clocks brought in phase, by applying the reset to all devices at the same time. The primary clock and the hard reset can be propagated to the device with minimal skew between devices, at least when using the GBTx, as long as the length of the optical fibres connected to the GBTx are of equal length. With their propagation delay of about 5 ns/m, they are the largest contributor to the skew. Trace lengths on the front-end boards also has a minor impact, but this can normally be compensated for when designing the boards. As the reset happens synchronously to the clock, there is the option to compensate for the skew by applying the reset signal one or more cycle earlier or later to certain boards, which gives a skew compensation granularity of 3.125 ns at the highest data rate. If the skew in the optical fibre length is not known to a high degree, the device provides the option[3] to

[3]This option is provided by the Bypass register, see Sect. B.1.9.

internally route one of the trigger inputs to the serial output. On the readout system it is then possible to count the number of cycles it takes from a pulse is sent to it is received again and by halving this, the length of the propagation delay can be found to within one clock cycle.

3.2.2 *Slow Control and Configuration*

The slow control of the SAMPA is managed through I^{2}C [6], using the 10-bit addressing scheme with automatic address increment on continued read/write operations. The design is based on an Intellectual Property (IP) block created by CERN.

The I^{2}C interface is designed to operate from 100 kHz to 5 MHz, which covers the most common speeds listed in the standard; standard mode (100 kHz), fast mode (400 kHz) and fast mode plus (1 MHz). Due to the need to operate at different speeds, the internal state machine is relying on the clocking supplied by the I^{2}C master to advance its state. In case something goes wrong during the transmission, there is the potential that the state machine will get stuck in an intermediary state. A watchdog is implemented that will time out and return the state machine to idle if the time between two falling edges of the I^{2}C clock line is more than 512 bunch crossing clock cycles apart (generally about 12.5 µs). This in sense sets the lower operating speed of the I^{2}C for the device. As the I^{2}C state machine operates at the bunch crossing clock frequency, nominally about 40 MHz, the design does not need to halt the communication to complete internal read/write tasks, even at the highest transfer speed, and so does not implement any clock stretching. The communication protocol is shown in Figs. 3.8 and 3.9 with the description in Table 3.1.

Another more conventional approach is to use the 7-bit addressing mode of the standard, which is more commonly used. However, for simplicity, the 10-bit addressing mode has been used instead, due to the way the implementation of the I^{2}C IP had been done. The I^{2}C standard only specifies how to read and write the devices, but does not specify how the access should take place if a device has multiple accessible

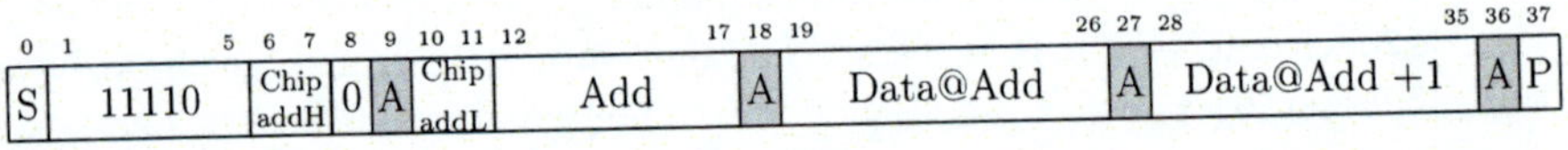

Fig. 3.8 Format for writing to the SAMPA. Gray boxes are bits sent by the SAMPA. See Table 3.1 for description of fields

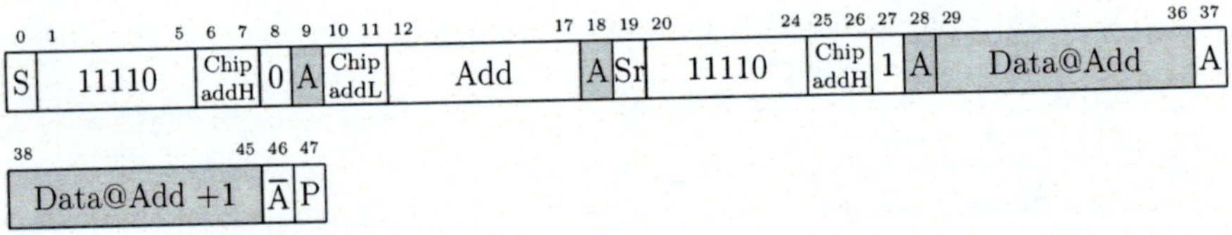

Fig. 3.9 Format for reading from the SAMPA. Gray boxes are bits sent by the SAMPA. See Table 3.1 for description of fields

Table 3.1 Protocol bit field descriptions of I^2C

Name	Bits	Description
S	1	Start
Sr	1	Start repeat
A	1	Acknowledge
$\bar{A}$	1	Not acknowledge
P	1	Stop
11110	5	Fixed preamble address for 10 bit addressing
Chip addH	2	Chip address [3:2]
Chip addL	2	Chip address [1:0]
Add	6	Register address
Data	8	Register data to be read/written

registers internally. The way that this is implemented in the IP is that parts of the device address space is used as internal register space, i.e. some of the upper bits are used for device addressing, while the lower bits are used for register addressing. The device addressing space needed is given by the MCH requirements, where up to five front-end cards, each with two SAMPAs, can be present on the same bus. This requires four bits of device addressing. Using 7-bit addressing only leaves three bits for register addressing, while using 10-bit addressing leaves six bits. As the S-ALTRO uses only nine directly accessible registers, the 64 available registers should be sufficient.

Another common way to solve the problem, like what can be found in the standard 24-series EEPROMs[4] [7], is to first transmit the address and afterwards the data for a write transaction. For a read transaction, the address to be written would be first sent to the device as a write transaction and then subsequently a read transaction is sent to read from that address. This method is though not available in the used IP and as the IP had been previously chip-proven, there was a reluctance to do architectural modifications to the design. However, the efficiency of using this method compared to the 10-bit addressing method is the same, as in the same number of words are needed for a read or write transaction.

The device is implemented with a global register bank, accessible through a Wishbone bus from the I^2C, which contain 40 addressable 8-bit registers holding configuration settings shared by the whole device, this includes registers controlling event management, daisy chain configuration etc. Each additionally channel has their own set of registers for configuration of filter parameters. These register banks have 31 addressable registers that are up to 13 bits wide. Compared to the previous ALTRO and S-ALTRO devices, this provides the opportunity to configure individually all

[4]Electrically Erasable Programmable Read-Only Memory.

parameters of each channel for the optimal filtering. In the previous devices, only a limited set of parameters were available for configuration on a per channel basis.

Access to the channel registers is done through the global register via a data input, data output, address, and operational control register. The information in the control register indicates which channel to access and if the operation is a read or write action. The order of the registers in the global register has been designed to minimize the amount of writes needed to update a channel register. As seen in Table 3.2, the address in the channel register address[5] to access is followed by the write data. which is again followed by the access control register. By utilizing the multi-byte write feature of the I^2C, a channel register can be updated in one continuous I^2C transaction. To write the same information to all channels, a broadcast bit can be enabled to broadcast the write action to all channels.

The device contains a 1024-word memory per channel, called the pedestal memory, which is primarily used for data corrections, further discussed in Sect. 3.2.3.1. The read and write port for the memories are accessible through the channel register, but the address port for the memory is shared among all channels and is accessible from the global register. To minimize the amount of writes needed to update the memories an auto-increment feature has been added for the pedestal memory address so that on each write to the access control register the pedestal memory address can be automatically increased without needing to update two more registers. Combined with the broadcasting of a channel register write to all channels, a full pedestal memory can be updated in 5 k 8-bit transactions corresponding to about 48 ms at 1 MHz.

For rerouting of input channels to specific serial links, the device contains a programmable register for defining the ordering, called the channel order register, further discussed in Sect. 3.2.6.1. The register is directly accessible through the global register via lookup. To save one address, only one address is used for the data during both read and write access. Like for the channel registers, the registers used for accessing the channel order registers are arranged to support a quick update.

Further information on the registers can be found in Appendix B.

3.2.3 *Digital Signal Conditioning*

To increase the efficiency of the lossy data compression algorithms and to facilitate easier analysis of the data during the final event-reconstruction, the sample data passes through some Digital Signal Conditioning (DSC). The DSC implements, in a pipelined manner, different algorithms to condition and shape the signal, and is comprised of four main building blocks

Baseline Correction 1 (BC1) This first filtering stage's main task is to remove low frequency perturbations and systematic effects. Different modes of operation are available, depending on the application.

---

[5]CHRGADD.

Table 3.2 Registers used for accessing the channel registers, channel ordering registers and pedestal memories. The order of the registers has been arranged to minimize the amount of writes needed to update a channel register

Register name		Address	Type	Default	Description	
PMADDL	[7:0]	0x15	RW	0x00	[7:0]	Pedestal memory address, lower byte
PMADDH	[1:0]	0x16	RW	0x00	[1:0]	Pedestal memory address, upper byte
CHRGADD	[4:0]	0x17	RW	0x00	[4:0]	Channel register address
CHRGWDATL	[7:0]	0x18	RW	0x00	[7:0]	Channel register write data, lower byte
CHRGWDATH	[4:0]	0x19	RW	0x00	[4:0]	Channel register write data, upper byte
CHRGCTL	[7:0]	0x1A	RW	0x00		Channel register control
			RW	0x00	[4:0]	Channel number
			RW	0x00	[5]	Broadcast to all channels (channel number ignored)
			RW	0x00	[6]	Write, not read from register address (returns to read after write)
			RW	0x00	[7]	Increment PMADD (returns automatically to zero)
CHRGRDATL	[7:0]	0x1B	R	0x00	[7:0]	Channel register read data, lower byte
CHRGRDATH	[4:0]	0x1C	R	0x00	[4:0]	Channel register read data, upper byte
CHORDAT	[4:0]	0x1D	RW	0x00	[4:0]	Channel readout order data
CHORDCTL	[5:0]	0x1E	RW	0x00		Channel readout order control
			RW	0x00	[4:0]	Position in order
			RW	0x00	[5]	Write enable

Digital Shaper (DS) This filter can be used for three different applications; tail cancellation, peaking time correction, or as a general purpose low-/high-/band-pass or band-stop filter. It is implemented as fourth-order Infinite Impulse Response (IIR) filter, specifically tailored for the two first applications, but can be reconfigured as a general second-order IIR filter for general purpose filtering.

Baseline Correction 2 (BC2) This unit applies a baseline correction based on a moving average filter. It can remove non-systematic perturbations of the baseline that are superimposed on to the signal.

Baseline Correction 3 (BC3) This unit applies a baseline correction through a slope based filter. It is provided as a more stable alternative to the BC2 filter.

3.2.3.1 First Baseline Correction

Time projection chambers, due to their inherent construction, suffer from a few signal effects that can be corrected through application of some filtering. A low-frequency spurious signal can be present that is in the order of a kilohertz or less. This signal perturbs the detector signal by shifting its baseline by an amount that is close to constant, i.e. a change of less than one ADC count during an event. This type of signal perturbation could, for instance, be caused by environmental changes e.g. drift in temperature and voltage, environmental variations due to differences in placement in the detector or by manufacturing or process variations.

Another type of signal perturbations is caused by systematic effects, like those related to triggering of the detector, which affect the signal in terms of superimposed noise patterns.

To cope with the first effect, a constant value can be used, in case the variation is very small, or an IIR filter can be used if the change is larger. The IIR filter is further referred to as the Variable Pedestal (VPD) filter.

The analogue front-end is constructed to add a small positive offset to the baseline of the input signal of about 100 ADC counts. Since process variations, temperature and supply voltage can impact, by shifting the baseline wither positively or negatively, the added offset provides enough margin so that input baseline of the signal should not appear below 0, which would render the data from channel unusable. Since this shift varies from channel to channel and chip to chip, there is an option to set a constant value per channel. The value to be used can be determined by temporarily enabling one of the other baseline correction filters, which can provide the calculated baseline through a registry entry. If the detector environment is stable and does not suffer from other signal perturbation effects, the constant value subtraction might be enough, which would enable the detector to save some power by disabling the other filters.

The VPD is best used if the baseline is close to constant within one event, but changes slightly in-between events, like what is shown in Fig. 3.10a. Since the VPD filter can be configured so that it only updates its calculated baseline outside an event, where there should be no detector signals, it will be effective for these cases. It will

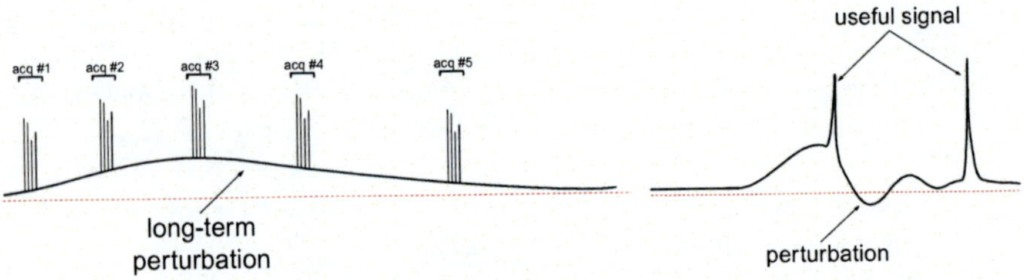

(a) *Example of long-term perturbations* (b) *Example of gating perturbations*

Fig. 3.10 Slow perturbation examples

then operate like the constant subtraction within an event, but with a freshly updated baseline value.

The VPD filter can also be configured to be always enabled and will then operate similar to the other baseline correction filters. To avoid inclusion of pulses in the baseline calculation, the filter uses thresholds, where if the signal is above or below the set thresholds, the input value is not included in the baseline calculation. To avoid that the signal gets stuck permanently outside the thresholds an auto-reset feature is available that will reset the filter if the signal stays outside the thresholds for too long. This resetting feature is similar to the one used for the Baseline Correction 2 (BC2), as will be further discussed in Sect. 3.2.3.3.

The VPD filter operates with a programmable response time of $\tau = 2^n$ sampling periods, where n is a value between zero and fifteen which needs to be adjusted to fit the noise of the environment. In sense, the baseline is computed as cumulative average of a programmable number of samples of the input signal.

$$y[n] = u[n] - \frac{x[n-1]}{\tau} \tag{3.1a}$$

$$x[n] = x[n-1] + y[n] \tag{3.1b}$$

Where $y[n]$ is the output value with the baseline removed, $x[n]$ is the cumulative sum, $u[n]$ is the input sample value and, $\tau = 2^n$ is the response time and n in τ is the number of samples the sum is calculated over.

If there are systematic effects, i.e. a fixed superimposed pattern is always present at a fixed delay from the start of an event, this can instead be corrected with the use of a pattern memory. The memory is programmed beforehand with the shape of the systematic perturbation and the shape is subtracted from the input signal for each event in the form of $y[n] = u[n] - m[n]$, where $y[n]$ is the input, $u[n]$ is the input and $m[n]$ is the memory value. The memory can subtract values up to the full programmable length of an event, which last a maximum of 1024 sampling cycles. A counter value, which increases at each sample cycle, is used for addressing the memory.

By using the memory as a look-up table, it is possible to, for example, perform non-linear conversion or to equalize the response across different channels. In this mode the circuit can perform a static conversion of the input signal of the type $y[n] = F(u[n])$, where the input value $u[n]$ is used as for addressing the memory. The output value $y[n]$ is stored in the memory position that is addressed. At any cycle n, the output $y[n]$ depends at most on the input sample $u[n]$ at the same time, but not on past or future samples of the input.

For testing purposes, the memory can also be used to inject a pattern into the data chain to allow testing of downstream logic, without the need to provide an external analogue signal in the front-end.

New to the SAMPA is the possibility to record the input signal directly to the pattern memory, this avoids the time-intensive process of acquiring the pattern to the offline system and re-uploading the data to the individual channel memories. New is

Table 3.3 BC1 baseline correction and subtraction modes. Legend: din : data input (samples); f(t) : pattern data; FPD : fixed pedestal data; VPD : variable pedestal data; f(din) : converted data

Modes of operation		Main configurations			
		din - FPD	din - f(t)	(din - VPD) - FPD	(din - VPD) - f(t)
Subtraction mode	Fixed	x		x	
	Time-dependent		x		x
	Variable			x	x
Conversion mode					
Test mode					
Modes of operation		Main configurations			
		f(din) - FPD	f(din - VPD) - FPD	f(t) - FPD	
Subtraction mode	Fixed	x	x	x	
	Time-dependent				
	Variable		x		
Conversion mode		x	x		
Test mode				x	

also that the 10-bit data stored in the memory can, at the output of the memory, be shifted to adapt to the 13-bit data in the data pipeline. It can either be 9 bits plus a sign bit, 10 bits, 9 bits plus 1 bit precision or 8 bits plus 2 bits precision. This provides the possibility to enter a more precise baseline value than previously where it was fixed at 10 bits.

The usages previously mentioned allow for three different modes of operation: subtraction mode, conversion mode and test mode. Some of these modes of operation can be combined allowing numerous configurations of the Baseline Correction 1 (BC1) circuit. While the fixed-mode and time-dependent-mode are exclusive, any of them can be combined with the self-calibrated mode as shown in Table 3.3.

The memory can be accessed through the I^2C, for both read and write, as long as the filter is not in a mode where the memory is being used in the processing chain.

In general, for this filter and the others discussed in this thesis, the adders that are used have been altered from how they were used in the previous devices to saturate instead of rolling over when overflowing. This should provide a more accurate value in extreme cases.

3.2.3.2 Digital Shaper

If a pulse arrives shortly after another one, the latter pulse will have an artificially higher peak as the pulse is overlaying on the tail of the previous pulse. This is particularly an issue for Multi-Wire Proportional Chamber (MWPC) where the tail can be quite long. The purpose of the digital shaper is to shorten the signal pulse and reduce the effect of pulse pileup. The architecture of the digital shaper is implemented as cascade of four first order direct-form 2 transposed IIR filters as shown in the top part of Fig. 3.11. Each stage of the filter is controlled by means of two coefficients, the zeroes L_i and poles K_i, that adjust the passband of the filter. The coefficients are programmed independently and have a precision of 13 bits. In general, the tail of the signal is shortened with increasing value of the poles, and shortened with higher value of the zeroes. This implementation corresponds to the following function in the Z domain:

$$H(z) = \frac{1 - L_1 z^{-1}}{1 - K_1 z^{-1}} \cdot \frac{1 - L_2 z^{-1}}{1 - K_2 z^{-1}} \cdot \frac{L_{30} - L_3 z^{-1}}{1 - K_3 z^{-1}} \cdot \frac{1 - L_4 z^{-1}}{1 - K_4 z^{-1}} \begin{cases} S = 0 \ \forall \ \ 0 \le K_i,\ L_i < 1 \\ S = 1 \ \forall -1 \le K_i,\ L_i < 1 \end{cases} \tag{3.2}$$

Where S selects if the coefficients K_i and L_i are signed. This is a new feature for the SAMPA and provides more freedom in the use of the filter, but sacrifices one bit of precision. In previous implementations, the coefficients were always positive.

If the coefficient L_{30} is set to 1 and $S = 0$, the filter has the same form as what was present in the S-ALTRO. This was an improvement in the form of less conversion noise, less power consumption and better accuracy then what was present in the ALTRO [8], which also only uses only three cascades.

Since GEM detectors do not suffer from long tails, the filter is not needed in the new ALICE TPC, but is kept as an option for other MWPC based TPCs. As the cascaded form is best aimed at removing signal tails [9], the filter has been modified to transform the two last cascades into a 2nd order direct form 2 transposed filter which can better provide more common filtering options like low-pass, high-pass and band-pass filtering capability. By setting the coefficients in the two first cascades two zero, the filter is reduced to only the 2nd order. The transfer function for the modified filter is shown in (3.3).

$$H(z) = \frac{1 - L_1 z^{-1}}{1 - K_1 z^{-1}} \cdot \frac{1 - L_2 z^{-1}}{1 - K_2 z^{-1}} \cdot \frac{L_{30} + L_3 z^{-1} + K_4 z^{-2}}{1 + K_3 z^{-1} + L_4 z^{-1}} \begin{cases} S = 0 \ \forall \ \ 0 \le K_i,\ L_i < 1 \\ S = 1 \ \forall -1 \le K_i,\ L_i < 1 \end{cases} \tag{3.3}$$

3.2.3.3 Baseline Correction 2

A second level of baseline correction can be applied to the signal to correct for signal perturbations created by non-systematic effects. The correction is based on a low-pass Finite Impulse Response (FIR) filter implemented as a moving average filter using

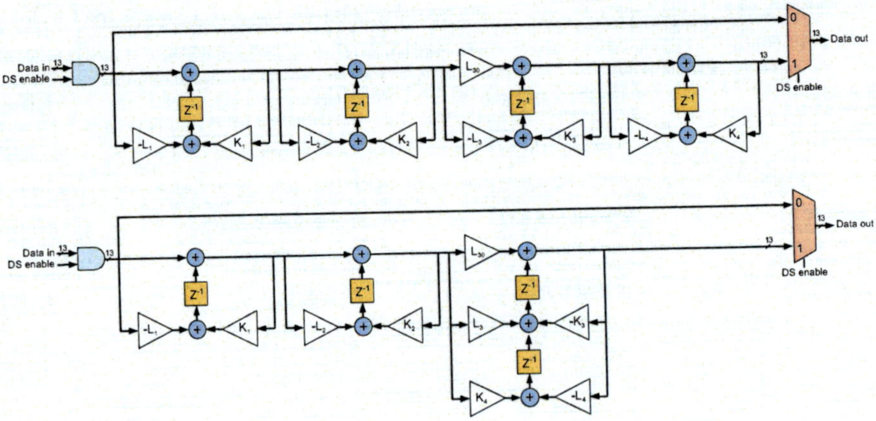

Fig. 3.11 Schematic overview of the IIR filter for the digital shaper

accumulated sum. The filter uses acceptance thresholds to exclude pulses from the average calculation, as inclusion of pulses would introduce undershoot in the output signal of the filter after a pulse, effectively introducing a dead time in the signal. When the signal is outside an acceptance threshold, either the higher threshold or the lower, the calculation of the average is frozen until the input signal again returns inside the thresholds (see Fig. 3.12a).

Inside the acceptance window, the correction is done according to the following equation:

$$x[n] = x[n-1] + u[n] - u[n-M] \qquad M = 2, 4 \; or \; 8 \qquad (3.4a)$$

$$y[n] = u[n] - \frac{x[n]}{M} \qquad M = 2, 4 \; or \; 8 \qquad (3.4b)$$

Where $u[n]$ is the input sample, $x[n]$ is the sum of the M last samples and $y[n]$ is the corrected sample value. When the current sample is inside the acceptance window, the average that is subtracted from a given sample is the average of this sample and the previous 1, 3 or 7 (depending on the configuration) samples that were inside the thresholds.

When a fast variation in the input signal is present, e.g. a pulse, the input signal will pass out of the acceptance window set by the thresholds and subsequently be excluded from the calculation of the baseline. When outside the acceptance thresholds, the value of the sample is corrected with the last updated baseline value from when the previous last sample was inside the acceptance window.

A setting to additionally exclude a programmable number of samples before and after any pulse that passes out of the exclusion thresholds is available. This might be useful if a slight undershoot is always expected to follow a pulse, or if the thresholds have been set rather high so there actually is a few samples that is part of the pulse before and after it passes the thresholds.

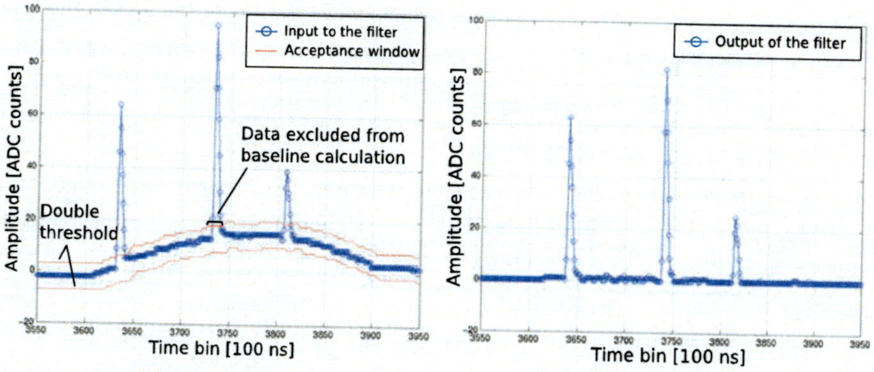

(a) *Pre-correction* (b) *Post-correction*

Fig. 3.12 Moving average principle

The glitch filtering implemented for the GdSP has been kept. The glitch filter, when enabled, prevents the activation of the post- and pre-samples when a single sample is outside the threshold. This makes it possible to keep the thresholds closer to the noise, which will provide more precision for the pulse exclusion.

3.2.3.4 Filter Changes and Improvements

As reported in [9], the ALICE TPC had previously experienced issues with the BC2 filter of the ALTRO, an example of this is shown in Fig. 3.13. In certain cases where the baseline is shifting down and up again, like seen at around time-bin 650, and a pulse is riding on top of the baseline change, the slope change might be steeper than what the filter can handle and the input signal settles at a level that is above the upper threshold for updating the baseline. Consequently, the baseline will not be updated and all the values on the output of the filter will be larger than zero.

The zero suppression compression that follows the filtering stages has the task of suppressing all values below a certain threshold value before storage, to save bandwidth and buffer space. The threshold level is usually set at zero, or at a few counts above, and everything below this level is suppressed and run length encoded. When all the values are above zero, all the samples are considered as one large cluster and every sample of the event is stored in the buffer. This quickly fills up all the available space, as the output bandwidth is not normally designed to handle the equivalent of a raw data readout.

The ALTRO has no way of resetting the filter, except by resetting the complete device, which means that the configuration registers needs to be reprogrammed as well. Like in the S-ALTRO, the SAMPA device also incorporates a soft reset functionality that resets the device, but keeps the configuration registers intact. To use the soft reset to recover from a stuck baseline, the upstream device would need to monitor the buffer overflow flag in the header of the data packets. As this is not an

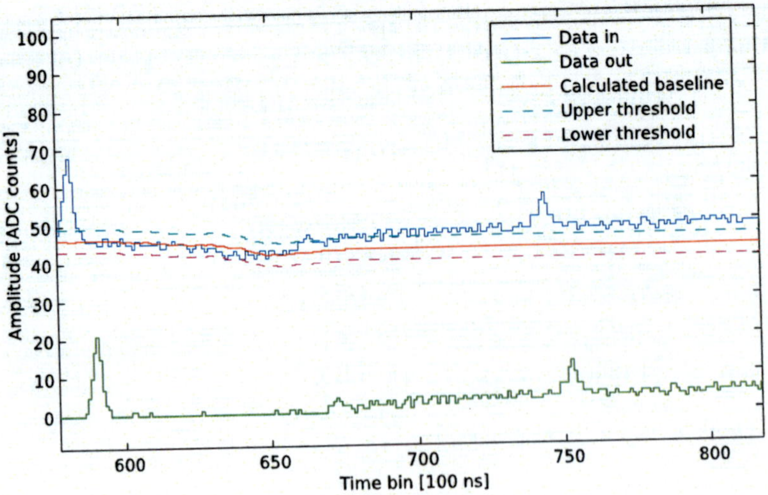

Fig. 3.13 Example of BC2 stuck baseline issue [9]

effective way of recovery, and as it requires that data is lost before it can recover, an auto-reset functionality is implemented in the SAMPA. If the baseline is found to be outside the thresholds for more than a given number of samples[6] the filter either reset the baseline to a pre-programmed value configured through the registry, or to the same baseline that the BC3 filter has calculated, as the BC3 filter does not get stuck. The GdSP has a separate reset for the BC2 baseline controlled from the configuration register, but this method has the same drawbacks as soft-resetting the device.

As an extra protection, an absolute threshold for the baseline, can be configured.

3.2.3.5 Baseline Correction 3

Conventional linear filters like the BC1 and BC2 have been used in the past in particle detector experiments and are useful for the task, but in some exceptional cases, they can end up in a failure mode wherein the baseline of the signal shifts to outside the thresholds of the filter and the filter stops to operate. In general, the filter has a max slope that can be handled before it needs to be reset or the thresholds need to be changed.

A new non-linear filter called BC3 [10] has been included, which does not have any thresholds. This makes the baseline tracking free of dead areas and the need for fine-tuning of configurations. The main idea behind this filter is to always follow the baseline, but with a limited slope configurable per channel. The slope parameter can be set individually for the up and down-slope.

The output signal $y[n]$ is expressed as a function of the input $u[n]$ as follow:

---

[6] Programmable from 0 to 1023 with a 4 count granularity.

$$y[n] = u[n] - x[n] \tag{3.5}$$

Where $x[n]$ is the tracked slope baseline value and can be expressed as:

$$x[n] = x[n-1] + \begin{cases} slope \uparrow & \text{for } u[n] > x[n-1] \\ 0 & \text{for } u[n] = x[n-1] \\ -slope \downarrow & \text{for } u[n] < x[n-1] \end{cases} \tag{3.6a}$$

$$x[0] = u[0] \quad \text{(arbitrary)} \tag{3.6b}$$

The filter has also a maximal slope, in the same sense as the BC2, but exceeding the maximum slope only causes a time-limited deviation. As the filter does not have acceptance thresholds, it also does not stay constant during an input pulse and as an effect, the output pulse will be slightly affected.

Figure 3.14 demonstrates the standard behaviour of the filter when changes of the baseline and different input pulses are applied.

When having large pulses, especially with longer shaping times, the upward slope of the BC3 filter will continue to rise during a pulse, until the pulse again passes below the current level of the calculated baseline. This will skew the waveform of the pulse and add noise. This can for instance be seen for the first pulse in Fig. 3.14, where the calculated baseline in blue rises, while the actual baseline has a slight downward slope. To oppose this, it would be possible to add acceptance thresholds for when the slope should stop increasing or decreasing, similar to the implementation on the BC2 filter. Though this will bring in the same issue as for the BC2 where the baseline can get stuck outside the thresholds. Therefore, this was not included in the current design.

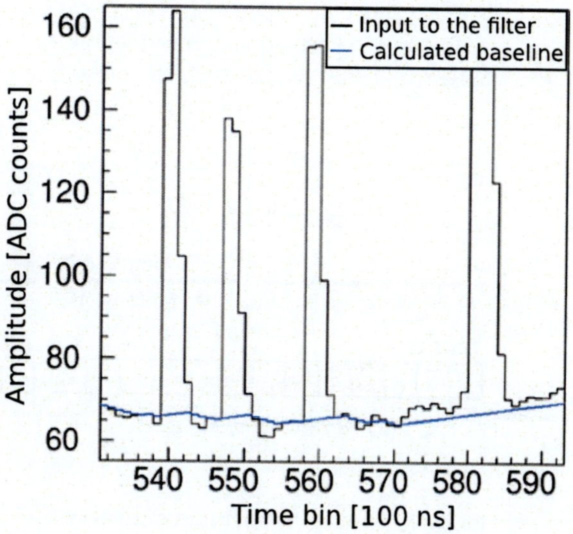

Fig. 3.14 BC3 filtering principle

3.2.4 Data Compression

Even though the SAMPA has enough output data links to handle reading out raw data, it is often not practical or economically feasible to do a raw readout. Different compression methods are provided to reduce the data amount enough so that fewer serial data links can be used. Both lossless and lossy compressions are available, depending on the detectors needs and resources.

Zero suppression This method removes all data below a given threshold, leaving only cluster data. As the data that is in between clusters are lost, it is a lossy encoding. This method is highly dependent on having a stable baseline to achieve good compression and minimal loss of information, so digital signal conditioning is normally needed.

Zero suppression with cluster sum This method operates on the same principle as zero suppression, but integrates the sample values in the cluster into one value. This compression method is suitable for detectors that have very clean signals and where only the time for the start of the cluster and the area is of interest.

Modified Huffman This compression method is a lossless compression, i.e. no information is lost. It uses differential encoding of the data combined with a lookup table to reduce the length of each word. This method is useful for detectors that want all data, but has limited bandwidth available. Depending on the detector data, it has a compression factor close to or better than zero suppression.

3.2.4.1 Zero Suppression

The device uses a pulse detection scheme with fixed thresholding, i.e. samples of value smaller than a constant decision level (threshold) are rejected. When a sample is found above this threshold, it is considered the start of a pulse (Fig. 3.15a).

In order to keep enough information for further feature extraction, the complete pulse shape must be recorded. Therefore, a sequence of samples (pre-samples) before

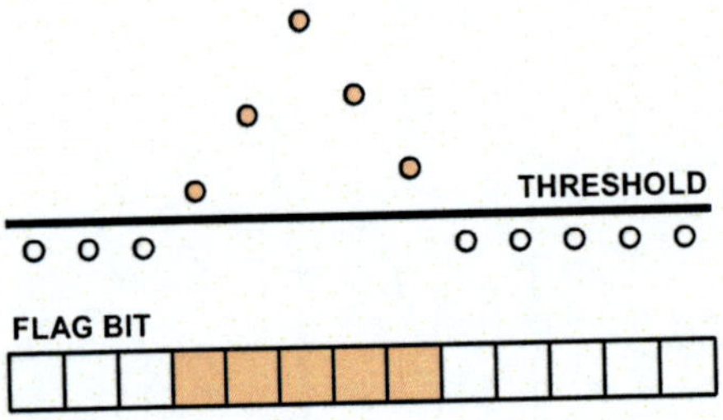

(a) *Application of threshold*

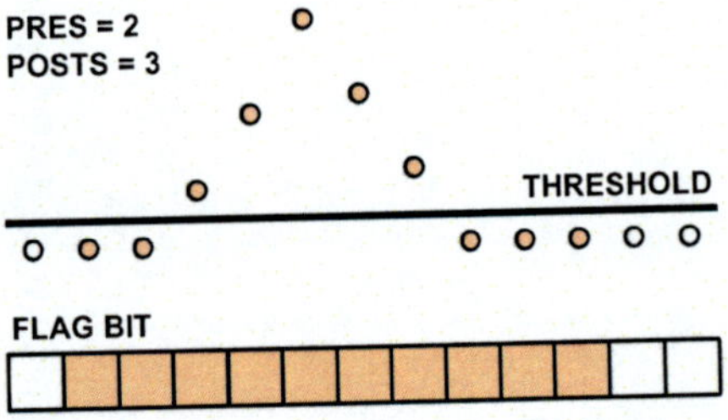

(b) *Feature extraction with two extra samples before pulse and three after*

Fig. 3.15 Zero suppression basic detection scheme

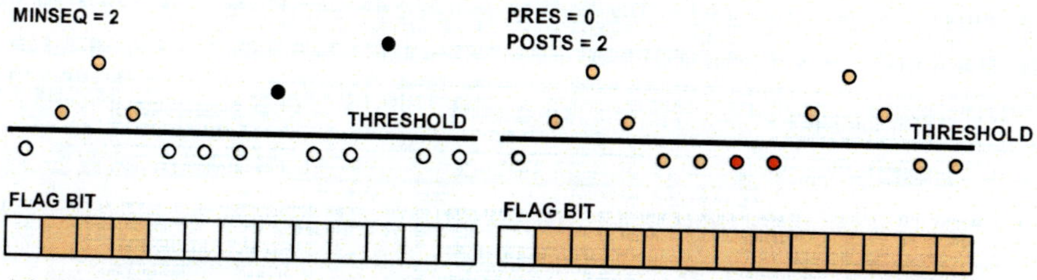

(a) *Glitch filtering with minimum samples above threshold of 2. Samples in solid black are treated as if they were below the threshold.* (b) *Merging of close clusters. Samples in red are included to make one complete cluster.*

Fig. 3.16 Zero suppression filtering and merging

the signal overcomes the threshold and a sequence of samples (post-samples) after the signal returns below the threshold are also recorded (Fig. 3.15b). The number of pre-samples and the number of post-samples can vary independently in the range between 0 and 3 for pre-samples and 0 to 7 for post-samples.

To reduce the impulsive noise sensitivity, a glitch filter checks for a consecutive number of samples above threshold, confirming the existence of a real pulse (Fig. 3.16a). The minimum sequence of samples above the threshold (glitch filter) which defines a pulse can vary from 1 to 3, not including any pre- or post-samples.

The pulse thus identified and isolated must be tagged with a time stamp, in order to be synchronized with the trigger decision for validation. Otherwise, the timing information would be lost by the removal of a variable number of samples between accepted pulses. This requires the addition of a time data to the set of sample data. Besides that, in a data format where the addition of flag bits is not allowed, a further word is needed to distinguish the sample data from the time data. This extra word represents the number of words in the set. Since for each new set of data we have two extra words, the merging of two consecutive sets, which are closer than 3 samples, is performed (Fig. 3.16b).

In case of the cluster sum mode, the data in one cluster is summed together into a 20-bit word and the requirement for the minimum distance between clusters is then changed to 3.

3.2.4.2 Lossy data compression

Due to the removal of a variable number of samples between accepted clusters in the lossy encodings (zero suppression, cluster sum), the timing information would be lost in the process. This requires the addition of a time-data to each accepted set of samples. Since 1023 is the maximum length of the data stream that can be processed by the SAMPA chip, the time information can be encoded in a 10-bit

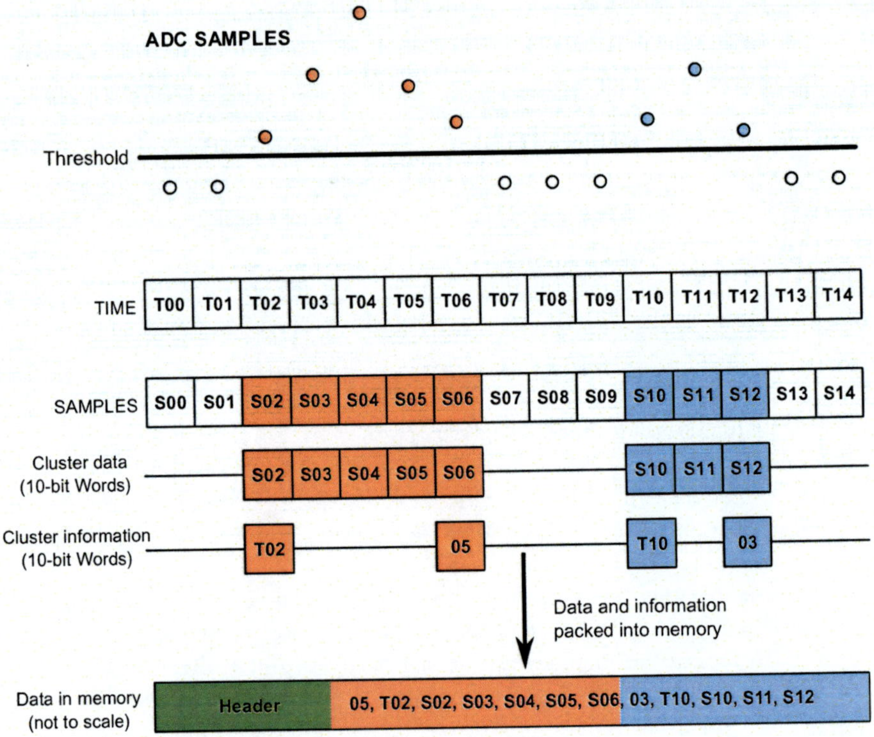

Fig. 3.17 The SAMPA data format for zero suppression encoding

word. The principle is to label each sample with a time-stamp that defines the time distance from the trigger signal. The time information added to each cluster during the formatting phase corresponds to the time-stamp of the first sample in the cluster.

The SAMPA data format does not make use of extra flag bits to distinguish the samples data from the time-data, but introduces a further word for each accepted cluster, which represents the number of samples in the cluster. In the ALTRO, the cluster size data included also the time information data and cluster information data in the count of the total number of words in the cluster, the SAMPA does not count them into the total.

These new 10-bit words, time data and number of samples per cluster, are introduced at the beginning of the cluster (Fig. 3.17). This is in contrast to how it operated in the ALTRO chip, where the information words were located after the cluster data [11]. This was done as it simplified the processing of the compression in their design since then the samples can be written first and the extra data is written in the next two available cycles until the next cluster starts.

The SAMPA overcomes this processing complication by having an early notification of when the cluster starts so that the time data can be written first and simultaneously a memory address skipped and reserved for later. When all the samples in the cluster have been written to memory and the number of samples in the cluster is known, it is possible to go back and insert this data at the front at the reserved address.

The principal drawback of the ALTRO method is that it uses extra memory and processing time on the computer that does the analysis. The complete packet needs to first be stored in memory as an array, if the source of the data is a streaming source, then it needs to be decoded back to front into another array so that it can be processed from the start of the packet. With the SAMPA method, the data can be processed directly from the streaming source and so the analysis can be completed faster.

The reason for using a time data to indicate where the start of a cluster appears instead of using the number of suppressed zeros is mainly due to historical reasons. In the ALTRO the time data gave the time stamp for the last sample in the cluster. This was easy to insert at the end of the cluster by keeping a global counter that counted time cycles from the start of the event. If there is a needed for compressing the zero suppressed data further with Huffman coding to save space for archival purposes, the number of suppressed zeroes will be easier to compress since it has a more narrow distribution, while the time-data would have a more flat distribution.

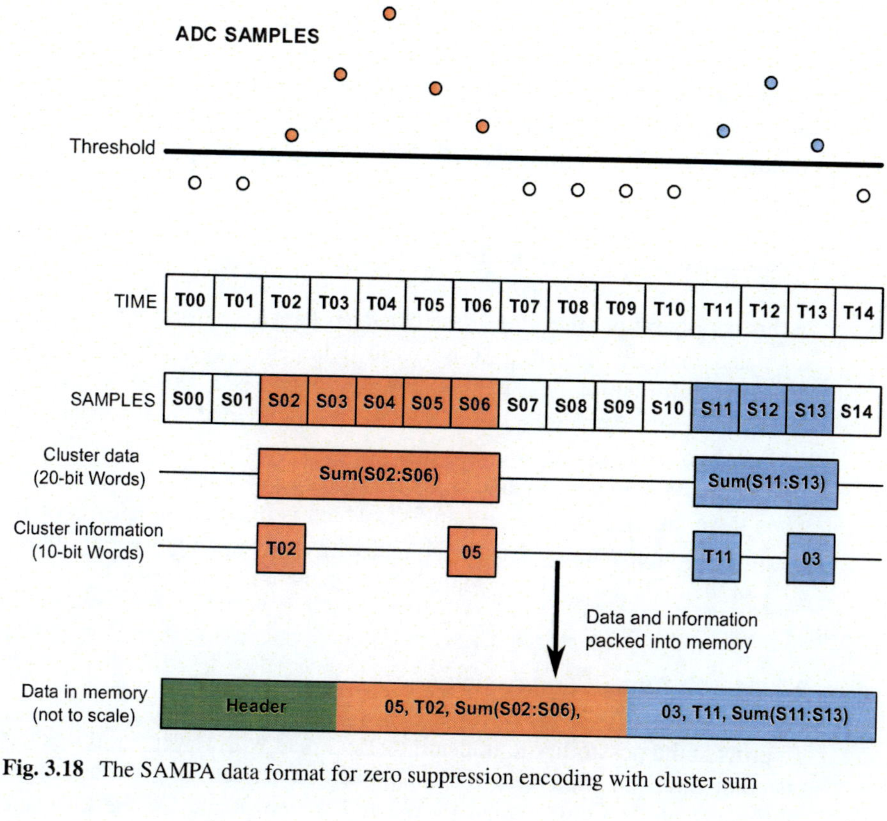

Fig. 3.18 The SAMPA data format for zero suppression encoding with cluster sum

3.2.4.3 Alternative Lossy Data Compression

For an efficient use of the sparse available bandwidth in daisy-chained mode, it is important to reduce the amount of data sent per cluster. However, as discussed in Sect. 2.4.3.3, the implemented cluster sum compression, as seen in Fig. 3.18, is sufficient for the MCH detector and so the following two methods for reduction of the amount of data sent are only proposed, but they have not been implemented.

Zero suppression with combined cluster sum and length

An alternative to the normal cluster sum mode is to combine the cluster sum and the cluster length into a 20-bit word. This can be done by constraining the size of a cluster to 31 samples, which will make it fit into 5 bits, leaving 15 bits for the cluster sum. The cluster sum can normally not overflow, as $31 \cdot 2^{10} < 2^{15}$, but the cluster size can possibly overflow, as the maximum size of a cluster is potentially 2^{10} if the zero suppression level is not configured properly. However, clusters can be artificially broken up into multiple clusters, if the length of the cluster is more than 31. As the time stamp for the start of a cluster and the length is present in the data, it is possible to join clusters together again when doing analysis. Since the average length of a cluster with the present longest shaping time of 300 ns and highest sampling rate of 20 MHz is about 20, the potential for overflow should be low, as long as the zero suppression level is set correctly. With this method, the total compressed size of a cluster would go down from four 10-bit words to three in optimal conditions, so the reduction in payload size would be 25 %.

Zero suppression with variable length sum

In this configuration, the sum of the cluster is kept in a 10-bit word when the sum is smaller than 2^{10}, and in a 20-bit word if larger. This is encoded in one bit of the length information, and so the allowable length of a cluster will be limited to 9 bits. Optimally this method can reach the same payload reduction of 25 % as the previously described method, it is less complex to implement, uses less area, and will not reach a bigger payload size than the normal cluster sum mode, except for when there are clusters longer than 511, which is uncommon. However, it is quite sensitive to the average pulse height, and with an average pulse height higher than 20 % of the ADC range there would be little benefit to this method.

3.2.4.4 Huffman Compression

To provide the possibility to transmit lossless data with fewer serial links, Huffman encoding has been implemented as was specified in Sect. 2.4.3.1. The length-limited Huffman table is implemented by reusing the constants for the tail cancellation filter of four channels. As the intention of Huffman is to transmit lossless data, the tail cancellation should not be enabled. The tail cancellation filter has nine 13-

bit constants, providing the possibility to have 36 Huffman values of length 12-bits, which was deemed sufficient and could provide a compression better than 2.5 required by the TPC for their initial specification [12]. If more constants are needed, it would be better to add a dedicated register table for the Huffman, but as it was not guaranteed to be used, this method avoids increasing the device area by having more registers.

3.2.5 Event Management

The event management's primary task is to notify each module of the data channel part when to enable processing of input data. An event lasts from when a trigger is received, until a programmable number of cycles, referred to as a time window, have been completed. The event management supports receiving the trigger to start the event, either from an externally pulsed pin, a command sent through the slow control interface, or by generating it internally whenever the current event is completed, commonly referred to as continuous operational mode. The device supports suppressing a given number of samples in the beginning and at the end of an event, which could be useful if the detector knows there is unwanted signalling at the start or at the end of an event. Suppression at the end can also enable detectors that normally would need to run in triggered mode, like for instance MWPC based detectors due to their need for gating to suppress ion-backflow, to use the continuous mode as the unwanted data during the gating is removed from the transmitted event.

As there are delays between a sample arriving at the different modules of the data pipeline, the event signals to each module in the pipeline must also be delayed. The current implementation is based on sending the event signals through a pipeline with the same length as in the data pipeline, with the signal for each module tapped at the appropriate delay. In the previous devices this was done by using a counter that was started at the start of an event and activated each module at the specific count. Any trigger that was closer than 20 cycles from the previous would not be detected. When the device is running in continuous mode, there is generally the need to synchronize the start of acquisition across all devices in the detector. The synchronization is done by sending an external trigger pulse to all devices at the same time. The devices then stop the current event and start a new one. As this could happen within the first 20 cycles of an existing event, the device needs to support spacing of triggers down to one cycle, which is solved by sending the signals through a pipeline. In case a new trigger is received during an event, this is marked in the header of the packet for the current event to indicate to the reviving system that the event was terminated early. The event management also captures the bunch crossing number for the time when the trigger was received and puts it in a pipeline for inclusion in the header.

A drawback of the pipe-lining method is that it requires more resources, particularly in case of the 20-bit bunch crossing counter. An improvement on the current implementation would be to delay all control signals to the bunch crossing counter module by the length of the pipeline, instead of delaying the values themselves.

3.2.5.1 Ring Buffer

The ring buffers that hold the event data is, as previously mentioned, split in two, where one memory holds the payload for an event, while the other holds the header. Due to the varying size of an event payload and the limited buffer space to hold the event payloads, this presents the possibility to voluntarily drop the payload data in case the buffer for payload becomes full, while retaining the header, possibly freeing enough space to store the next event payload. In case of lost payload, the header will be altered to mark that the payload was dropped. As long as the occurrence of overflow is low, this should provide more complete payloads then if the packet was terminated whenever the buffer got full.

The header is only written to the buffer once the event is completed. The writing to the header buffer, in turn, signals to the buffer readout section that it can start reading out a new event. In this way, the readout part never reads out from the data buffer until an event is completed and so it is safe to move the pointers in the data memory about. The number of samples to read from the payload memory is indicated in the header.

Since the buffer, due to the new implementation of the zero suppression, needs to sometimes write two words to the buffer in one sample cycle, the payload buffer writing always operates at half the serial clock cycle to give the possibility to write two words in one sample cycle, regardless of the sample speed. In hindsight this could probably have operated at the bunch crossing clock speed, since the maximum sampling clock speed is 20 MHz which is half the bunch crossing clock speed and this in turn would save a little bit of power consumption. With some optimization it would also have been possible to provide a second clock domain at two times the sampling speed, though this would require some extra work in the back-end due to the new clock domain. As the module that receives the data from the output of the buffer also operates at half the serialization speed, the synchronization logic between the input and output could in principle have been greatly simplified, but since the change to the faster clock domain was changed late in the development process, this was not done. For SAMPA v1 there was a buffer in between the module that sent data to the buffer and the buffer itself, but this in-between-buffer had the potential for overflow if the data rate was higher than the sample rate, even though the output serialization could handle the higher flow.

The addresses for accessing the buffer are Gray encoded which should reduce the switching in the memory when it is writing in sequential access. The buffers are implemented close to what is described in [13]. Another technique as described in [14] was originally used for v1 of the device, but was deemed too risqué to be used for further development due to its asynchronous method of passing pointers for buffer full and empty checking.

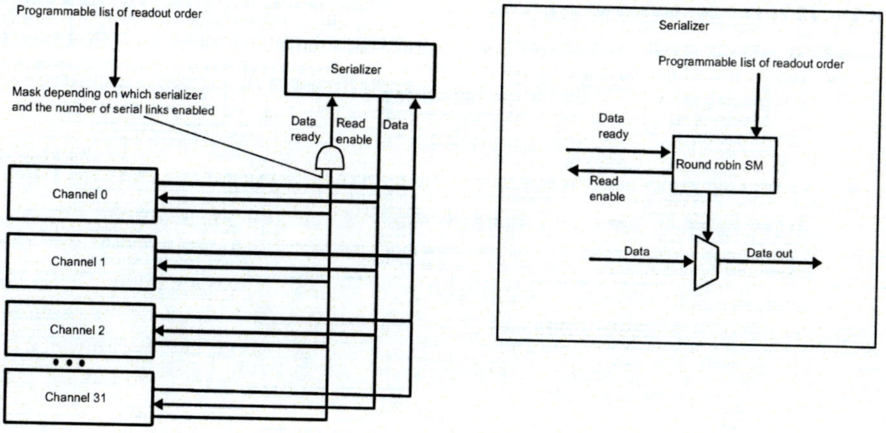

Fig. 3.19 Conceptual overview of the serial readout implementation

3.2.6 *Readout*

The device supports two primary ways of data readout, a packet-based serialized readout and a parallel readout. The parallel readout bypasses all the data processing of the device and pushes the raw ADC data directly out on the data lines, while the packet based method enables all the processing of the device and can send the data out on a programmable number of links, up to eleven. In the packet based mode, the data can so be sent from one to another device in a daisy-chained fashion, to lower the number of upstream serial links that are required.

3.2.6.1 Serialized Data Readout

The serialization of data is as sketched in Fig. 3.19. The drawing shows the readout for a single serial link of the available eleven. The serializer has no knowledge of format or content of the data it serializes. It receives a data-ready signal from a channel that indicates data is ready to be read out and it sends a signal back to the module to tell it to advance the data it presents to the serializer. It will read from the selected channel for as long as the data ready signal is high. A masking of the data ready signal is done to restrict the number of channels a serializer can read data from, depending on the number of serial links that have been enabled. The number of channels per links is divided equally and the serializer reads data from each channel in a round robin fashion and operates independently of each other. A priority is enabled so that if a heartbeat packet needs to be transmitted, it always has the highest priority and will be sent after the current packet is completed.

When the serializers have no data to send, i.e. there are no heartbeat packets or channel data, they will transmit sync packets to keep the receiving end in sync. The sync packets are the same size and format as a packet header, but the information in it is fixed. As this keeps the links toggling while the device is idle, it will consume some extra power. An option would have been to power off the serial links whenever there are no more data to send and when a new packet is ready to be transmitted, a sync packet would be sent first to enable the reviving side to synchronize to the data stream. Currently power-off is only done for a serial link when it is not enabled.

DC balancing of the output serial links is not implemented in the device, as it is not a requirement for interfacing with the GBTx. In other applications that do not use the GBTx, the DC balancing might be of benefit to provide better signal integrity. As the native word format of the SAMPA is 10-bit based, the best option would be to use a 5b/6b encoding which would increase the overall bandwidth usage by 20 %. The existing sync packets would not be necessary any more as the DC balancing code has unused codes that can be used for link synchronization.

To reduce cost and development time, it is common to produce only one or maybe a couple of different versions of a front-end card to mate with the detector pads. However, the detector might have several different mappings of physical pads to input channel on the front-end card, depending on the position in the detector. As the serial links are independent of each other, the data from one chip can potentially be sent to two or more upstream devices for further processing of the data for instance for event reconstruction. If it is important to the detector to have specific input pads arrive at specific upstream devices, independent of the mapping, the SAMPA provides a way of remapping which input channels are presented to each serializer.

Due to the setup with a fixed set of channels available to each serializer, the SAMPA currently provides no way of load balancing between the serializer in case a set of channels for one serializer has a higher occupancy than the set for another. This could cause unnecessary overflow of buffers, even though there is available bandwidth on other serial links. If all serial links would have access to read from all of the channels, it will provide a more equal load across the serial links. Each link would then iterate through its subset and whenever all the channels in its subset is empty, it picks another one not in its subset and which is non-empty and not already being read. Normally it would send sync packets in this case. This method will however cause the predictability of which channel will arrive next on which link to be lost. It will also render the option to re-order which channels connect to which serial link mostly pointless.

3.2.6.2 Daisy Chained Readout

For detectors with very low data rates, specifically the MCH, a daisy chaining option is implemented. This lets multiple devices share a single serial link, increasing the number of devices that can be read out by a single GBTx. Each device is provided with an extra data input port and two data-control signals for this purpose. A connection diagram can be seen in Fig. 3.20.

The upstream device is set up to run with one serial downlink and the busy out signal from the downstream device is connected to the busy in of the upstream device.

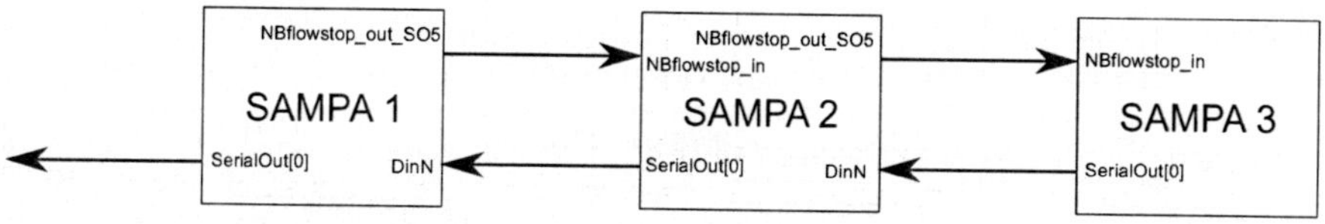

Fig. 3.20 Connection setup for daisy chaining

When the busy signal is high, it tells the upstream device to halt its transmission of data after the current packet is done. While waiting to send data again the upstream device will send sync packets to keep the communication in sync.

The downstream device has the serial output of the upstream device connected to its daisy input port. Sync packets arriving on the link are filtered out and only heartbeat packets and data packets are forwarded. The accepted packets from the upstream device are added to a ring buffer in the same way as for the internal channels. A configuration register defines how many upstream devices there are from the current device. This value is used by the serialization module that serializes packets from the ring buffers for defining the number of packets that should be forwarded from the upstream link before a packet from one of the internal channels can be forwarded. The handshaking between the serial link module and the module that handles the readout of the buffers is however constructed in a way that it is not possible to read out two consecutive packets from the same buffer without a few cycles delay, which causes there to be sent a sync packet in between two packets from the upstream device when this setting is enabled. In principle there is no limiting factor to fix it in the existing buffer readout module, but there would need to be an addition of an extra state machine in the buffer readout module that pre-reads the next header from memory so that it is ready to be transmitted. The primary issue however lies in that the data ready signal from the buffer also is used to indicate that a stream of data is completed and that the serializer should select another buffer or a sync packet to send, but this could be solved by separating the data ready and packet done signal. Since the MCH will only operate with two devices in a chain, this was not prioritized to be fixed for the SAMPA v3.

Commonly, all the devices in a chain will be clocked from the same source. As the data is sent and captured at the same clock speed, it is possible that the data stream could be corrupted after exiting the synchronizer on the receiving end if the data that is received changes too close to the clock edge causing metastability in the synchronizer. Since the time between data being clocked out of the upstream device to it is received on the downstream device is dependent on various factors such as the distance between the devices, the properties of the Printed Circuit Board (PCB) material, the process variation of the device etc. a programmable delay chain has been added before the signal enters the synchronizer. The incoming data can be delayed by up to 12.5 ns in steps of 0.2 ns. The maximum length allows for delaying the signal by a full clock cycle at 80 MHz and provides enough of a granularity to allow for 15 steps of delay at 320 MHz.

The busy signal is not provided with a delay chain as the signal is level based. A delay in the reception of the busy signal of one or two cycles due to metastability in the synchronizer is not considered a problem.

To recover from communication errors during transmission between the devices, the receiving unit does Hamming correction on the received headers. Errors are corrected before storing in ring buffer for retransmission. In case of the detection of double errors, the receiving unit drops the data and waits for resynchronization. The receiving unit can also detect a stuck high/low data input.

When the daisy chaining option is not in use, the units relating to the daisy chaining is turned off, through clock gating, to save power.

A drawback of this readout method is that if there is a malfunction of one of the devices in a chain that makes it inoperable, there will also be no data received from the previous devices in the chain. It is also vulnerable to channels or devices that are overproducing data, either through misconfiguration or from malfunctioning input channels. Options are available to turn off misbehaving channels, but monitoring must be in place upstream to detect and reconfigure the device. In case a device is partly operational, it is possible to configure it so that the complete device is bypassed by forwarding the control and data signal directly from input to output.

3.2.6.3 Direct Readout—Serialization

For detectors that would prefer not to use the data handling capabilities of the SAMPA, a mode is available where the raw ADC samples are directly serialized and the rest of the digital circuitry is powered down through clock gating. This mode operates with a serialization speed of 32 times the ADC sampling speed and can be configured in two modes, referred to as the normal mode and the split mode.

In the normal mode, the 10-bit data for channel 0 will be output in parallel on the serial links. In the consecutive cycle, the data for channel 1 will be put out, and so on. Since the serialization speed is 32 times the ADC sampling speed, it is possible to transmit the current sample for all of the 32 channels in the time it takes to sample the next sample.

In the split mode, five first serial links are dedicated to channel 0-15 and the other five serial links are dedicated to channel 16-31. The serialization speed is still 32 times the ADC sampling speed, but it takes two serialization cycles to transmit a full sample. In the first cycle, the 5 lower bits for channel 0 and the 5 lower bits for channel 16 will be transmitted. On the consecutive cycle, the 5 upper bits for the same channels are transmitted, it then continues with the 5 lower bits for the next channel, and so on. In this way, it is possible to direct the data for half the channels to one upstream receiver and the other half to another.

Upon start up, a 32-cycle sync pattern is transmitted, so that the receiving end can synchronize to the stream. As the data transmission is cyclic, there is no need to have a separate data bit to indicate the start of a new sample. If the receiver loses track, it can restart the transmission to get a new sync word without losing more than a couple of samples.

The control circuitry for the serialization is protected from Single Event Upsets (SEUs), as further discussed in Sect. 3.4.1, but the internal generation of the ADC sampling clock is not. This can be problematic for detectors that need to have the sampling clock in phase across several devices. To mitigate this, the spare 11th serial link is used to transmit the ADC clock together with the data stream. An SEU in one of the ADC clock derivation registers will present itself as a phase shift in the clock. By counting the number of high and low cycles on the receiving end of the 11th link, it is possible to determine if an SEU has occurred and from there it can be considered if the device will need to be reset.

This serialization method has the benefit that it is simple to implement a receiver for it on the upstream device. It also opens the possibility for implementing other compression forms and filtering methods that are more suited to the detector application, with a trade-off in added development time. The filtering and compression can be done separately on the front-end card by use of an FPGA. However, it has the drawback of increased power consumption of the front-end card and the added effort of radiation qualification of the FPGA device. It can also be done off-board by sending the data through the GBTx to a readout device. The drawback then will be that there will be an increase in cost in the form of a need for extra GBTx devices and optical fibre links. Since there is no error correction added in the data itself and there is no synchronization during the data transmission, it is up to the upstream device to verify the integrity of the link during the initial 32-cycle sync pattern. As all serial links will toggle in a predetermined pattern, it can be verified that none of the links are stuck at a fixed value.

3.2.6.4 Direct Readout—Combinatorial

For applications where there is only a need to use a subset of the channels the device provides, a mode is available where the data from the ADC channel inputs are multiplexed to the 10 serial link outputs. The channel that should be used is determined through the configuration of five inputs pins. Data from several channels can be acquired by cycling through input pin configurations during a sampling cycle. As the configuration pins for setting the channel are single-ended CMOS, the cycling speed would be limited to the single-ended switching speed of the driving device. Since the data rate using this method is much lower than with the other methods, it opens up the possibility to interface the device to low cost FPGA or a microcontroller solutions.

3.3 Design for Test

The production, manufacturing, and packaging of ASICs, as well as the soldering of packaged devices on to printed circuit boards, are all imperfect processes where there is a probability of ending up with a device that does not operate according to specifications. Therefore, there is a need to verify the functionality, preferably after completing each step, so that you end up with a fully functional system. Catching the bad devices earlier in the process can save on costs, as fewer devices would need to be processed in the following steps.

Due to the large number of devices that will be needed for the final production of the SAMPA, and due to the increase in complexity of the device compared to its predecessors, it is not feasible to test the device fully in a reasonable time frame only by verifying the functionality through operating the device and changing its configuration. Since the device also employs various error correcting techniques to combat SEUs, as further discussed in Sect. 3.4, errors from manufacturing defects might end up being masked by the error correcting, but will still affect the SEU tolerance of the device. Due to this, the design has been implemented with several design-for-test features.[7]

To detect errors in manufacturing of the ASIC, the design is implemented with scan chain capability. When a certain configuration is enabled, all registers in the design are stringed together into five chains. By manipulating input signals, shifting bits through the chain, and analysing the output signals, it is possible to exercise most of the logic in the design and thereby verifying that there are no broken connections or elements.

The memories in the device are made by a third party and do not support any scan-chain capability. Instead, a dedicated test structure has been created to verify their functionality.

To verify that the input/output pads of the ASIC has been properly bonded to the pins of the package, and to verify there is a proper connection between the package pins and the printed circuit board once the device is soldered-on, a Joint Test Action Group (JTAG) boundary scan functionality has been implemented. Through a simple standardized interface, it is then possible to exercise digital output pins of the device and read digital input pins.

The analogue section and ADC are not implemented with any built-in test functionality; it therefore requires a functioning digital section for verification. The testing is primarily limited to observing that the baseline of the channel is within expected limits, that the channel responds to input signals and that the pulse gain is as expected from the setting and provided input signal.

3.3.1 Scan Chain

In a scan-enabled design, the registers (flip-flops) are connected together in one or more scan chains. This enables enable access to the internal nodes of the design from outside. By shifting patterns into the input of the chain, the state of the design is set up in a particular way. By advancing the state of the system by one clock cycle, the

[7] When features are added to a design with the primary purpose of verifying the functionality of the device, it is commonly referred to as design-for-test.

Table 3.4 Scan chain information

Design version	Non-scannable flip-flops	Non-scannable latches	Scannable flip-flops	Flip-flops part of a scan chain	Coverage
2	3519 (2.8 %)	1544 (1.2 %)	121 411 (96 %)	121 295 (95.9 %)	91 %
3	2646 (2.1 %)	72 (0.1 %)	123 026 (97.8 %)	122 980 (97.8 %)	92 %

value of the next state is captured in the chain. This state can then be shifted out and compared against the expected one. By varying the state that is set up, most of the in-between logic functionality can be verified. As the state space is very big, it is difficult to reach complete coverage. The coverage of the design for the SAMPA v2 and v3 is given in Table 3.4.

The scan chain functionality cannot determine that the overall functionality of the circuit is correct. It can only verify that the design has been assembled correctly according to the recipe that was provided, i.e. that all the logic gates are in place, operating and connected properly. Functional testing that verifies the device is operating according to specification is further discussed in Chap. 4.

Since the rearrangement of the design into a scan-enabled design and the creation of the vectors for the testing is a complex process, the process is automated by a tool. In this case, it was done with Cadence Encounter Automatic Test Pattern Generator.

Testing of devices can normally be done after the manufacturing of the dies in the manufacturing plant by use of commercial automated test equipment. It is also normally done after packaging by using custom testing equipment. Doing the testing before packaging has the potential to save costs by avoiding to package defective devices. However, as the device needs to be tested post-packaging as well, to verify that he packaging went well, the cost saving might not be there if the yield of the manufacturing is high enough, since a large portion of the devices would then need to be tested twice. For the SAMPA, all devices have been packaged to lower production costs.

The choice of how many chains to have is a compromise between the time needed for testing and the number of input/output pins dedicated to the scan chain. Fewer and longer chains means clocking data through the chain will take more time during testing. As the SAMPA is somewhat pin-limited, a choice of five chains was selected, which gives about 25 k registers per chain. To reduce the number of pins dedicated to the scan chain implementation, the pins for the output of each of the chains are reused as serial data output pins during normal operation.

The non-scannable flip-flops and the flip-flops that were not integrated as part of the chain, as given in Table 3.4, are primarily related to the clock divider, reset circuitry and the memory Built-In Self-Test (BIST). A separate test-enable pin is used for disabling the excluded circuitry during testing. The clock and reset circuitry are instead exercised through the control of the external clock and reset pins during the scan test. The memory is exercised in its own test as discussed in the next section.

In a worst-case situation where, after manufacturing, all of, or a large portion of, the devices are non-operational, it is possible to utilize the scan chain capability for debugging the issue. The scan chain can be used for dumping the state of the device as well as injecting values to alter the state. As this was not the case for the SAMPA, this functionality was not utilized.

3.3.1.1 Scan Observability

The fault coverage of the design could potentially have been higher if more adaptions to the design would have been done earlier in the design process to help increase the observability of unreachable parts. Sections of the design, like the JTAG, BIST, memory, and clock generator have been excluded as they interfered with the controllability of the chain. Additionally, there are pins that go to and from the analogue section which are not controllable. Assuming that the tool that generates the coverage vectors is able to reach 100 % coverage of the logic that it can observe, the total coverage would still only be 98.9 % in the version 3 design due to inactive logic and uncontrollability. To increase the observability for logic that surrounds black-boxed sections, it is possible to add what is referred to as shadow registers, which observe each input to the black box. These registers have no operational function when the design is operating normally, but when the design operates in scan chain mode they can observe the inputs to the black boxes. In the same way, it is possible to connect the output of a black-boxed module to a multiplexer together with an output from a shadow register. The path of the shadow register through the multiplexer is only enabled in the test mode and so, in this way, the output side is controllable as well. The same method can be applied to capture data going out and in from pins that are not accessible outside the device, like for instance the signals to and from the analogue section.

3.3.2 Memory Built-In Self-Test

The memories used in the SAMPA are dual and single port 8-transistor Static Random Access Memory (SRAM) IPs provided by ARM [15]. Memory IPs are used for their higher memory densities, lower power, ease of design and for their proven design. The IPs that are used do not provide BIST for testing of manufacturing defects, so a custom built-in memory test has been designed to provide quick pass/fail testing.

The test structure is built as a wrapper surrounding each IP, where the data, address, and control lines for the memories are multiplexed between the normal data path and the test structure. The test is initiated by pulling a pin high externally to the chip.

The algorithm used for testing is a 5N length test (3N for dual port memory), where N is the number of addresses. Meaning it takes this amount of clock cycles to complete a loop and completely verify a memory. The test will loop continuously as long as the test is enabled, providing the added possibility to also use the test for evaluating the memories' Single Event Upset (SEU) probability in a irradiation test.

The test is based on writing a fixed pattern, in this case a checkerboard pattern, of 0x2AA to the memory in the first iteration, then in the next iteration it will read and check the 0x2AA pattern and write back a pattern of 0x155 (the inverse pattern) and in the next it will check the 0x155 and write 0x2AA and so on [16]. The test algorithm is similar to the Modified Algorithm Test Sequence (MATS) march algorithm [17]. The algorithms are normally only designed for single port memories, but for testing of the dual port memories, it can be simplified, as they are in the current design always written on one side and read on the other. Thus, it is possible to do the read and write in the same cycle, though not at the same address, so the write is always lagging one address behind the read.

When an address in the memory is read out and it contains an incorrect value, the test structure will set a signal high for one clock cycle indicating an error. The signals indicating an error from each of the test structures are ORed together to provide a combined output signal. Externally to the device, there are two status outputs available, a latched single-ended output, and a pulsed differential output. The pulsed output will give one pulse per address where an error is found on, while the latched output will stay high until an error is found and will then stay low, enabling a slower test system to detect an error without the need to sample the pulsed output at a high speed. The latched output is primarily intended for production testing, while the pulsed output can for instance be used for SEU detection during radiation testing.

The minimum running time for one test iteration is determined by the largest memory that runs on the slowest clock, which is the pedestal memory. The decision on which test to implement is a trade-off between test time and fault coverage. The 5N chequerboard sequence that is implemented does not provide a full coverage of all possible errors, but is fast, requiring only 256 μs when operated with input clocks of 320 MHz and 20 MHz for the serial clock and ADC clock respectively.
The chequerboard tester implemented is primarily able to detect:

Stuck-at faults A stuck-at fault occurs when the value of a cell or line is always 0 (a stuck-at-0 fault) or always 1 (a stuck-at-1 fault).

Stuck-open faults The connection to a cell is broken returning the previous read value.

Address decoding faults Faults present in the address decoding logic. This includes failure to access a cell located at an address, shorts between addresses, stuck-at address lines, and stuck-open address lines.

However, it is not able to, or only partly able to detect:

Coupling faults A write to one cell changes the content of a second cell. This includes a transition in one cell that inverts another cell, a transition in one cell that forces a 0 or 1 in the other cell, and state coupling faults where a cell is only forced to a value if the coupling cell is in a certain state. Only a fraction of these can reliably be detected with the currently used test algorithm.

Transition faults A cell or line fails to undergo a transition from 0 to 1 or 1 to 0.

Data retention faults When the cell is unable to hold its value. In SRAM this could be caused by defective pull-ups. Data retention is tested by adding a delay after the normal sequence is done and then continue after the delay. As the SAMPA does not rely particularly on a long retention time in the memory, this was not added in the automatic test. On the other hand, it would be possible to test this by stopping the clocks to the SAMPA for a while before enabling them again.

Read disturb fault Reading the cell multiple times will flip its content. Multiple reads on the same cell are not done with the current algorithm.

A more common industry technique is to use a 13N algorithm as with the Inductive Fault Analysis (IFA) algorithm [18] or a 14N as with March SRD [16] for complete coverage. Full coverage for the previously mentioned techniques is only valid for memories with individual cell access. Word-oriented SRAM, where a complete word is read and written at once, introduces the problem of state coupling faults between two cells at one address. To detect these faults all four state transitions between two cells must be tested. This would be done by running the same test with different patterns. For a 10-bit memory like in the SAMPA, this would require five iterations of the algorithm [18], significantly increasing test time. Coupling coverage could have been a bit better if the true physical layout of the cells in a word would have been known so that the chequerboard pattern is represented in the actual layout.

For using the pulsed output from the device for SEU detection, there are some considerations;

- Multiple changed bits in the same address are only detected as one error.
- All the memory tests are run in parallel with the error outputs ORed together, so if two memory testers discover an error at the same time, they will mask each other.
- The pulse length for a found error is one clock cycle of the clock domain the memory tester is running at. If an error is found in two or more consecutive addresses, the error signal from those addresses will be conjoined.

All of this will add to the dead time of the system as well as the uncertainty in the total number of errors found.

A possible extension to the BIST is to make it controllable through the JTAG. This would provide the possibility to supply different patterns to the tester, which would cover testing of more coupling faults. By adding some extra complexity, it would be possible to make the algorithm programmable. The default simple test would then provide enough detection of faults before packaging, while the extended testing could be done post-packaging if a higher fault detection was required.

3.3.3 JTAG Boundary Scan

JTAG (IEEE/ANSI Standard 1149.1-2013) [19] defines a standard for adding a dedicated JTAG port to a device for debugging purposes. Internally the device runs a state machine, which operates independently of the rest of the device and is controllable through the serial interface of the JTAG port. Several devices can be chained together by connecting the serial data output of one device to the serial data input of the next. Instructions can be loaded into the device to control the behaviour of the state machine. The SAMPA supports the minimum required instructions, which are to sample the input pins, load the output pins and bypass the device.

There can be manufacturing errors at several steps in the production that can lead to inaccessible or shorted pins. There can be broken or shorted bond wires between the die and the carrier PCB of the package. There could be manufacturing errors in the carrier PCB of the package. There can also be bad solder connections between the package and the front-end card as well as broken traces on the front-end card. The JTAG provides a quick way to verify that there is a connection between an external device and the chip internal to the package and that there are no shorts between the signals. The same is possible with the scan chain capability as well, but the scan chain test takes longer to complete and will require analysis of the results to determine what the specific error is.

3.3.4 Ring Oscillator

A ring oscillator circuitry has been implemented to track process variations in the manufacturing and determine the effect that changes in supply voltage and temperature has on the speed of the digital design. The circuit diagram is shown in Fig. 3.21. It consists of the inverter ring, a frequency divide-by-16, two counters running respectively on the divide-by-16 ring oscillator clock and a reference 10 MHz clock and some glue circuitry to control and end the test. The length of the ring was chosen so that the operating frequency would be 160 MHz in the typical temperature/voltage corner. Consequently, the frequency in the worst-case corner would be 100 MHz, and 220 MHz for the best case, based on simulation and available propagation delay values from the manufacturer's datasheet. The number of stages is odd to sustain an oscillation and it is a prime number to reduce the likelihood of higher harmonic resonances being present [20]. As the enable signal is generated locally on-chip it will have a rather monotonically increasing sharp edge and have little noise, combined with the relatively low speed of the oscillation there should be little chance of injecting multiple pulses due to the transitioning of the enable signal from "0" to "1".

To be able to automatically determine the frequency of the ring oscillator, the frequency is divided by 16 and compared against the ADC clock. A counter for each clock is started at the same time and after the ADC clock counter has run for 255

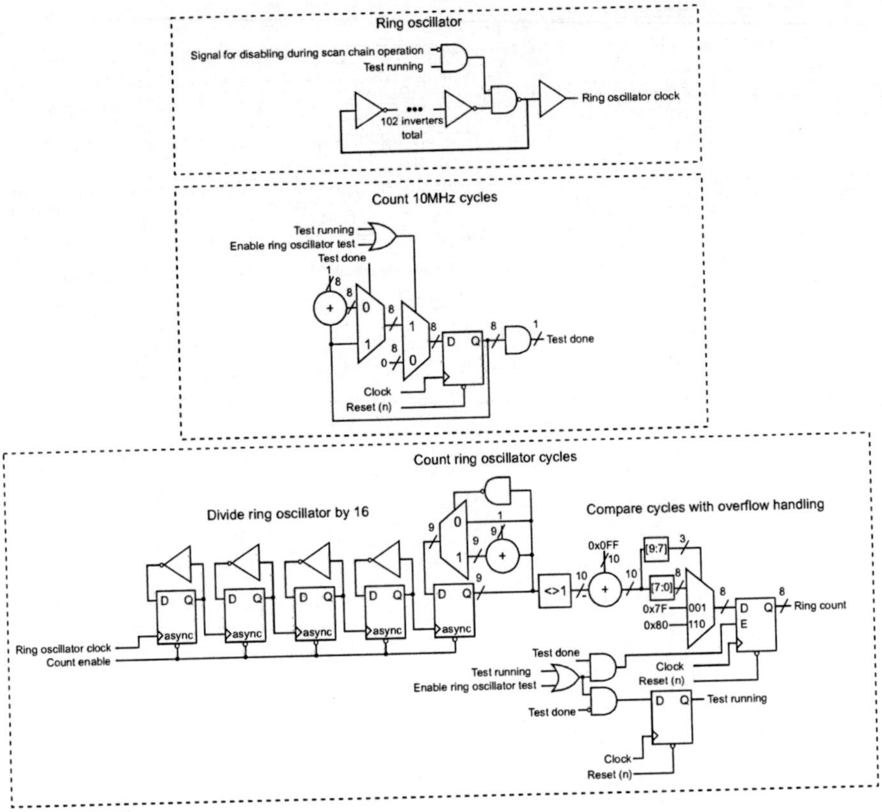

Fig. 3.21 Schematic of the ring oscillator circuitry

cycles, it will stop the ring oscillator counter. The difference is calculated, the result is checked for arithmetic overflow and is then stored in a register (RINGCNT) available through I²C. The RINGCNT value is a signed byte and the oscillation frequency can be calculated as

$$f = \frac{(255 - RINGCNT) \cdot 16 \cdot f_{ADC}}{255} \tag{3.7}$$

where f_{ADC} is the ADC clock frequency. When the test has completed, the oscillation in the ring will be disabled automatically. Through some multiplexing, the ring oscillator signal can be passed to one of the differential serial links for the duration of the test, so that the signal can be observed on a scope. During scan-chain testing the ring is disabled to prevent it accidentally being enabled during a test.

If an effort is done to calibrate the frequency at the nominal operational voltage versus the ambient temperature during operation, the ring oscillator could operate as a simple temperature sensor during detector operation. As the ring oscillator does not interfere with the regular operation, this can be done at regular intervals as part of the detector control and monitoring system.

3.3.5 Miscellaneous Test Features

Some other test features have been implemented to aid in the verification of parts of the design. A multiplexer is inserted before the output of serial link 0, which can output signals from internal circuitry, or from any of the inputs, depending on the setting of a configuration register. The signal from the either of the three trigger inputs can, for instance, be set up in loopback mode and routed back out. This can be used for determining round-trip delay of the system in a production environment. For instance, if the device that receives the data from the SAMPA is located far away, there could be a significant propagation delay of the data due to the length of the optical fibres or wires. By configuring the device in this loop-back mode, sending a pulse to one of the trigger ports and counting the cycles until a pulse is received back will enable a coarse time estimate to be determined. This loop-back mode can also be used for qualification the SLVS driver/receiver pair with the help of an external Bit Error Rate Tester (BERT).

The daisy chain data input can also be routed directly out so that the complete device can be bypassed when it is used in a daisy chain mode. This can be beneficial in case a device stops operating properly, but the data from the second device can still be used. The signal can also be tapped after the controllable input-delay compensation chain for the daisy chain data input, so that the actual delay can be measured.

A Pseudo Random Binary Sequence (PRBS) generator has been designed for easy qualification of the SLVS driver independently of the receiver. The sequence is generated from a 31-bit maximum length Linear Feedback Shift Register (LFSR) that feeds back the XNOR of output 31 and 28. The generator runs on the serial link clock for maximum speed.

A simple ADC serializer can be set up to serialize a single channel for test purposes. Selection of the channel is done by changing a configuration register (SERCHSEL). A 6-bit preamble (b010011) is sent MSB first and the sample data follows. The samples are serialized with the serial link half clock so that running with a main clock of 320 MHz and an ADC clock of 10 MHz there will be a total of 16 bits sent.

The internally generated clocks from the clock divider circuitry can be output to verify jitter and frequency stability. As the clock divider supplies a delayed clock to the digital design in relation to the clock that goes to the ADCs, it is possible to compare the intrinsic delay between the clocks and to quantify the delay for each step configuration of the programmable delay chain.

The clock generated from the ring oscillator described in Sect. 3.3.4 can be output to determine more accurately the frequency, or to view the waveform on an oscilloscope. The oscillation is only active after the ring oscillator test has been enabled, and then only for 256 ADC clock cycles or 512 oscillator cycles, whichever is the shortest. It should oscillate at about 100 MHz in the worst-case corner, 220 MHz for the best case and 160 MHz for the typical case, based on simulation and available propagation delay values from the manufacturer's datasheet.

3.4 Error Handling

With the increase in the interaction rate expected for Run 3, the radiation load on the SAMPA will consequently also increase. The highest radiation environment is present in the TPC where the high energy hadron flux is expected to reach about $3.4\,\mathrm{kHz/cm^2}$, compared to $0.8\,\mathrm{kHz/cm^2}$ in Run 1 [21]. High energy hadrons are the primary source of radiation-induced SEEs in the ALICE environment. The SEE issues of primary concern for the SAMPA digital design are Single Event Upset (SEU), where a particle can flip a bit in a memory element, and Single Event Transient (SET) where a single particle creates a transient on a line that can further propagate to a memory element if the event happens close to a clock edge. Both of these issues can be recovered from by correcting the data either manually or automatically.

Single Event Latch-up (SEL) issues can also occur, where the particle creates a temporary short between the supply voltage and ground and which needs to be recovered from by power cycling or lowering the supply voltage. This issue is though layout related and must be mitigated by altering the layout or by substituting problematic library components by non-affected components and therefore cannot be directly mitigated on the design level.

Devices in a radiation environment are also affected by dose-related issues that alter the analogue properties of a design over time, the more dose that is absorbed. As the expected dose rates for TPC in Run 3 is only 2.1 krad [21], this is not considered an issue, as at this level, without considering annealing, the change in analogue parameters will be low [22].

Errors induced by SEE can be, for the SAMPA, classified according to three severity levels: data-path errors, configuration errors, and functional errors. Data-path errors are the least severe errors as the errors occur in parts of the circuitry that are periodically overwritten. The errors are commonly confined to within one event. Configuration errors are errors that occur in the configuration registers or pedestal memory and which will alter the operation of the device. These errors will remain until the device is reset or reconfigured, and so these errors are of a more severe character. It is, however, possible for the device to operate normally with these errors, depending on which configurations changed. Functional errors are the most severe as they can prevent the device from operating properly. These errors might happen in the state machines, the memory pointers or in certain counters. In the SAMPA, all parts of the design are protected against SEE, except for the sections that are part of the data-path and test structures.

Further information on the radiation tolerance and susceptibility of the device can be found in [23].

3.4.1 Single Event Upset Handling

A common technique to increase tolerance to SEE is to create custom cells (i.e. flip-flops and latches) that have been improved by altering or adding extra components (resistors, transistors etc.) to mitigate the effect of the injected charge. The use of custom cells comes with an increase in size and power consumption, in addition to the extra work needed to design test and qualify them. The SAMPA design uses standard cells and as such need to mitigate the effect of SEEs through design instead.

SEUs are caused by charged particles traversing a device, injecting charge along the way and potentially causing a register's value to change to the opposite value. A common technique to protect devices against SEUs is by employing Triple-Modular Redundancy (TMR), which is essentially the use of a two-out-of-three voting concept at a low level [24]. As the information is stored in multiple locations and, provided that a hit only affects one node, the upset will not propagate and the correct data can be recovered. Accumulation of errors can be prevented by making sure that the register is refreshed periodically or automatically when an error is detected. Most of the operationally important registers in the design have been protected through TMR to avoid functional upsets in the device that would require the device to be reset. There are about 125 000 flip-flops in total in the second prototype design, including the extra TMR registers. The amount of registers (excluding extra TMR flip-flops) is about 55 000, of which 20 700 (38 %) are not protected by TMR.

The exceptions are:

Internal data path The number of registers in the data path between the ADC and the buffer memory amount to about 17 700 registers or about 1/3 of the design. SEUs in the data is deemed to be acceptable, so TMR is not enabled for these registers to save area.

Daisy chained data path The number of registers on the data path from the daisy-chained input to the buffer memory amounts to 109 registers. These are not protected for timing reasons and as the header-data is already protected through SECDED.

Test structures Registers in the test structures, i.e. the ring oscillator, LFSR generator, JTAG, and some other test infrastructure, are not protected since they are kept in reset by TMR protected configuration registers during normal operation. The test structures contain 138 registers.

Memory BIST The memory Built-In Self-Test is not protected as it is kept in reset by an external pin that is pulled low. It contains about 2700 registers.

Figure 3.22a shows the schematic drawing of a normal TMR register and Fig. 3.22b shows the schematic for a TMR protected synchronizer. The TSMC library has a custom component for the majority voter which reduces the additional area needed for a TMR register compared to a non-TMR register.

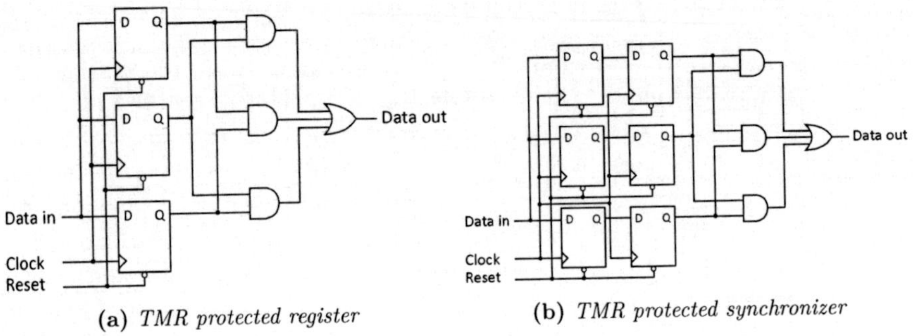

Fig. 3.22 TMR protection

```
`define TMRINST(HIGH,LOW,NAME) \
 wire [HIGH : LOW] NAME; \
 reg  [HIGH : LOW] NAME``_TMR[2:0]; \
 voting_nbit_3reg #(.SIZE((HIGH)-(LOW)+1)) \ ,
   NAME``_voter(.A(NAME``_TMR[0]), .B(NAME``_TMR[1]), \
                .C(NAME``_TMR[2]), .out(NAME))
`define TMR(NAME, RANGE, ASSIGNMENT) \
 { NAME``_TMR[0] RANGE, NAME``_TMR[1] RANGE, \
   NAME``_TMR[2] RANGE} <= {3{ASSIGNMENT}}
module tmr_example(clk, din, dout);
  input clk, din;
  output `TMRINST(2,0, dout);
  always @(posedge clk) begin
    `TMR(dout,[0], din);
    `TMR(dout,[2:0], dout[1:0]);
  end
endmodule
```

Listing 3.1 Macro implementation for TMR protected registers.

A common problem when implementing TMR through coding is that the synthesis tool, which converts the Register Transfer Level (RTL) description into gate level, will see the two extra registers as redundant and simplify them into a single register, effectively removing the TMR. This was solved in the current design by using a macro function to instantiate and assign values to the registers, as shown in Listing 3.1. The macro adds a suffix "_TMR" to the name of all TMR protected registers, the synthesis tool is then later told to preserve all registers with this suffix.

The assignment macro supports partial range assignments as well as assignments to the full range if the range is not given. Complementary "NOTMRINST" and "NOTMR" macros exist for registers without TMR so that they can be easily converted to TMR with minimal code changes. Having the TMR instantiation as macros also adds the possibility to disable completely TMR by redefining the macro. This can be used for speeding up simulations when it is deemed that the TMR will not influence the simulation. To avoid the accumulation of errors, it has been made sure that all registers have a feedback path from the voter output to the register input for the cases where the register does not update each clock cycle.

Another option that has been employed is to triplicate the modules containing the registers that should be protected. This was used for the global- and channel-configuration register modules, as all the registers needed protection and it made for cleaner code in the core module. The drawback of the method is that all registers need to exit the module to connect to the majority voter one level up, and then the voted signal needs to go down into the core modules again. This amounts to a lot of wiring and extra work connecting modules and signals together. It is also prone to mistakes if register widths are changed or registers are added or removed.

To avoid that the synthesis tool simplifies the modules into one, a suffix "_MODTMR" is added to the top module and the tool is told not to do any optimization/simplification in the module. The drawback is that this also prevents the tool from doing beneficial simplifications to the other logic in these modules. The SAMPA only uses this method for the global and channel register units, as they contain mostly registers. It is also applied to the I^2C slave module as the code was supplied in this form by CERN.

3.4.2 Single Event Transient Handling

In a TMR protected design, there is still a possibility of having SETs that alter the value of a signal interconnecting two registers. As this value will propagate to all three registers of the TMR, the value will be latched by them all, and the state of the system will have been changed to an incorrect one.

Normally SETs are mitigated by triplication of all the logic in the design, including the clock and reset nets. This comes in addition to the triplication of the registers, so there are effectively three separate versions of the design. Voting between the designs is done as normal after each register. As the probability of getting SETs are in general low, the amount of extra area that triplication of the full design would require was not considered justified.

The design includes preventative measures to at least avoid that SETs create functional errors by making sure all state machines return to an error state in case invalid combinations of input signals are provided to the state machine. Data taking for the channel will be halted until the next triggered time window is started.

3.4.3 Upsets in Data Memory

Since the design uses a separate memory for the header and payload data, and since the length of the payload for a given header is encoded in the header, it is important to protect the header memory against upsets. This is so that when it is time to read out a packet from memory, the correct amount of words is read out from the data memory. The header is therefore encoded with Hamming(50,43), including an extra parity bit for SECDED, which protect the pointer from single errors. The header is corrected after it is read out from the memory and if a double error is found the state machine resets its memory pointers to the start of the current packet that is being prepared on the write side. A parity bit of the payload is also included in the header, providing a way to determine for the readout system if there has been an upset in the payload data.

Due to space concerns, the design does not use any EDAC techniques on a per address basis for the memories as this would require one or more extra bits per address, increasing the design by 10% for a parity bit, 40% for Hamming(14,10) or 50% for SECDED with Hamming(15,10). A parity bit would only be of benefit if the specific data could be marked as bad and excluded from further processing. This would only be practical to do for the payload memory, but would require that the extra parity bit per payload word be transmitted along with the payload, significantly increasing the bandwidth needed.

The time-based subtraction memories (pedestal memories) and the memories used for the pre-trigger delay do not have any form of built-in mitigation added. Since the pre-trigger memory is essentially an extension of the sample data-path, which is also not protected, an error in this memory is of no great concern, as it will only alter the value of a single sample. An upset in the pedestal memory would, however, be persistent and of greater concern. Like for the data memories, there is no protection included here either due to limited device area. A technique to mitigate upsets in the pedestal memories is to refresh the content between data taking runs, or whenever an opportunity becomes available, where the memory is not in use.

3.5 Design for Low Power

Power consumption in a CMOS device consists of dynamic power when the logic is transitioning and leakage power when the logic is in a non-switching static mode. The leakage power is mainly dependent on the manufacturing process and the supply voltage. As the digital and analogue have separate power domains, it is possible to lower the digital supply voltage if a power reduction is needed. This would come at the cost of lowering the maximum operational speed and lowering the noise margin. Timing verification has been done to verify that the device operates correctly in the worst-case characterization corner, which has the voltage at 1.08 V. An option would have been to have several power domains in the digital design that could be independently powered down, but this would likely negatively affect the power routing, which possibly would lead to an increase in noise.

Dynamic power is due to switching of capacitances and the short circuits that occurs from supply to ground when transitioning from one output level to another. The switching power is proportional to $\alpha f C {V_{dd}}^2$, where α is the activity factor i.e. the fraction of the circuit that is switching, f is the clock frequency, C is the node capacitance and V_{dd} is the supply voltage. Therefore, the dynamic power can be lowered by reducing the operational clock frequency, preventing logic from switching when not used, disabling the clock to certain sections when not in use or by lowering or turning off the supply voltage to certain sections.

The majority of the design operates at the lowest necessary speed, which is the same as the ADC sampling speed. Only the circuitry between the ring buffers and the serializers, which is a small part of the design, runs at a higher speed, and then only at half the serialization speed. The parts that run at higher speed are also concentrated furthest away from the analogue section to reduce the noise coupling. As the digital part that runs at the ADC sampling speed will still create noise when the logic is switching, there is a possibility that this could influence the sampling of the analogue signal in the ADC. Therefore, a programmable delay chain has been implemented for the sampling clock that goes to the ADCs so that the sampling can happen during a quiet period. The delay chain used delay cells to implement the delay with a granularity of 1.5 ns per bit for a maximum of 94.5 ns. The delay in a delay cell is subject to process variations, temperature, and voltage variations, so if an accurate delay is needed for keeping several devices in sync, it is also possible to only invert the clock.

To avoid unnecessary toggling of data paths, the paths that are not in use are kept at static values e.g. the inputs to filters that have been disabled and the address and data ports for memories that are not currently activated.

Clock gating is used to disable the clock to circuitry that is not needed. For instance, when the daisy chain mode is disabled, the clocks for the module that receives data from the daisy chain are turned off. The I^2C module enables its clock only between a start bit and a stop bit, so it is effectively disabled when it is not receiving a word. A watchdog makes sure the module returns to idle if a start is received, but the transmission halts before a stop is received.

If a specific channel is determined to be defective or the device is used for an application where not all of the channels are in use, it is possible to clock gate the specific channel through a configuration register. This clock gating disables the clocks to the ADC and prevents the digital processing chain of the channel to run, though it does not clock gate the digital channel to avoid having 32 separate clock trees. As the digital channels consume most of the area of the device, having separate clock trees would likely negatively affect the routing of the device.

In the direct readout mode, only a very small part of the design is needed, primarily the direct serialization module, the slow control, and the main configuration register. Therefore, the clocks for the rest of the design are gated to save power. Enabling of the clock gating is done through an external pin, which also enables the direct readout mode.

Table 3.5 Power consumption on the digital rail at different operational frequencies

Serial clock (MHz)	ADC clock (MHz)	Voltage (V)	Switching (mW)	Leakage (mW)	Total (mW)
320	10	1.2	18.35	7.47	25.82
160	5	1.2	9.1	7.47	16.62

When the device is configured to run with less than the maximum number of serial links, the current source for the Scalable Low-Voltage Signalling (SLVS) differential drivers [25] of the unused links are disabled to save power. The drive strength of the used serial links can also be configured in three steps so that if the receiving device is located close to the SAMPA, the drive strength can be decreased to lower power.

An option would have been, for applications that do not need full TMR, to have partial or full disabling of the TMR circuitry to save power. However, due to the way the TMR was implemented, this could not be easily be implemented.

The power consumption of the device operating with different clock frequencies is listed in Table 3.5, it can be seen that the power scales linearly, mainly with the main operating ADC clock.

References

1. H. D. H. Herrera, "Noise and PSRR improvement technique for TPC readout front-end in CMOS technology," Ph.D. dissertation, Escola Politécnica, São Paulo, Brazil, 2015. [Online]. Available: http://www.teses.usp.br/teses/disponiveis/3/3140/tde-05072016-151016/.
2. G. Gramegna, P. O'Connor, P. Rehak, and S. Hart, "CMOS preamplifier for low-capacitance detectors," *Nuclear Instruments and Methods in Physics Research Section A: Accelerators, Spectrometers, Detectors and Associated Equipment*, vol. 390, no. 1, pp. 241–250, 1997. https://doi.org/10.1016/S0168-9002(97)00390-2.
3. C.-C. Liu, S.-J. Chang, G.-Y. Huang, and Y.-Z. Lin, "A 10-bit 50-MS/s SAR ADC with a monotonic capacitor switching procedure," *IEEE Journal of Solid-State Circuits*, vol. 45, no. 4, pp. 731–740, Apr 2010. https://doi.org/10.1109/jssc.2010.2042254.
4. M. Arora, *The Art of Hardware Architecture: Design Methods and Techniques for Digital Circuits*. New York, NY: Springer New York, 2012. [Online]. Available: https://doi.org/10.1007/978-1-4614-0397-5_2.
5. C. E. Cummings, "Asynchronous & synchronous reset design techniques - part deux," in *Synopsis User Group Meeting*, 2003.
6. NXP Semiconductors, *I2C-bus specification and user manual*, 6th ed., April 2014.
7. Microchip Technology Inc., *24LC16B Datasheet*, 2016. [Online]. Available: https://www.microchip.com/datasheet/24LC16B.
8. E. G. Garcia and R. E. Bosch, "Low power optimization of an IIR for tail cancellation in High Energy Physics applications," in *2009 Ph.D. Research in Microelectronics and Electronics*, July 2009, pp. 28–31. https://doi.org/10.1109/RME.2009.5201329.
9. T. Naaranoja, "Digital Signal Processing for Particle Detectors in Front-End Electronics," Master Thesis, University of Helsinki, Helsinki, Finland, 2014. [Online]. Available: http://hdl.handle.net/10138/135874.
10. K. Münning, "Impact of BC2 and BC3 on simulated data, comparison," SAMPA MPW2 design review II Meeting. [Online]. Available: https://indico.cern.ch/event/382908/contributions/1809834/attachments/763500/1047483/Slides-20150401-SAMPA-Filter-Comparison.pdf.
11. The ALICE TPC Collaboration, *ALICE TPC Readout Chip User Manual v. 0.2*, CERN, 2002. [Online]. Available: http://ep-ed-alice-tpc.web.cern.ch/ep-ed-alice-tpc/doc/ALTRO_CHIP/UserManual_draft_02.pdf.
12. S. Klewin, "Length-limited Huffman coding," TPC CRU Meeting, 2015. [Online]. Available: https://indico.cern.ch/event/453573/contributions/1951400/attachments/1169252/1687190/klewin_2015_10_13.pdf.
13. C. E. Cummings, "Simulation and synthesis techniques for asynchronous FIFO design," in *Synopsis User Group Meeting*, 2002.

14. C. E. Cummings, "Simulation and synthesis techniques for asynchronous FIFO design with asynchronous pointer comparisons," in *Synopsis User Group Meeting*, 2002.
15. ARM. (2017) Single port and dual port SRAM embedded memory IP. [Online]. Available: https://www.arm.com/products/physical-ip/embedded-memory-ip/sram.php.
16. A. J. van de Goor, "Using march tests to test SRAMs," *IEEE Design & Test of Computers*, vol. 10, no. 1, pp. 8–14, Mar 1993. https://doi.org/10.1109/54.199799.
17. R. Nair, "Comments on "an optimal algorithm for testing stuck-at faults in random access memories"," *IEEE Transactions on Computers*, vol. C-28, no. 3, pp. 258–261, March 1979. https://doi.org/10.1109/TC.1979.1675331.
18. R. Dekker, F. Beenker, and L. Thijssen, "A realistic fault model and test algorithms for static random access memories," *IEEE Transactions on Computer-Aided Design of Integrated Circuits and Systems*, vol. 9, no. 6, pp. 567–572, jun 1990. https://doi.org/10.1109/43.55188.
19. IEEE, *Standard for Test Access Port and Boundary-Scan Architecture*, IEEE Std 1149.1-2013 (Revision of IEEE Std 1149.1-2001), May 2013.
20. M. Bhushan and M. B. Ketchen, "Generation, elimination and utilization of harmonics in ring oscillators," in *2010 IEEE International Conference on Microelectronic Test Structures*, mar 2010, pp. 108–113. https://doi.org/10.1109/ICMTS.2010.5466847.
21. The ALICE Collaboration, "Upgrade of the ALICE read-out and trigger system," CERN, Tech. Design Rep. CERN-LHCC-2013-019, ALICE-TDR-015, 2013. [Online]. Available: http://cds.cern.ch/record/1603472.
22. L. Gonella, F. Faccio, M. Silvestri, S. Gerardin, D. Pantano, V. Re, M. Manghisoni, L. Ratti, and A. Ranieri, "Total ionizing dose effects in 130-nm commercial cmos technologies for hep experiments," *Nuclear Instruments and Methods in Physics Research Section A: Accelerators, Spectrometers, Detectors and Associated Equipment*, vol. 582, no. 3, pp. 750–754, 2007, VERTEX 2006. https://doi.org/10.1016/j.nima.2007.07.068.
23. S. M. Mahmood, "Exploring Single Event Effects in SAMPA ASIC for Time Projection Chamber (TPC) and Muon CHamber (MCH) detectors in A Large Ion Collider Experiment (ALICE)," Ph.D. dissertation, University of Oslo, Oslo, Norway, 2019, to be published.
24. R. E. Lyons and W. Vanderkulk, "The use of triple-modular redundancy to improve computer reliability," *IBM Journal of Research and Development*, vol. 6, no. 2, pp. 200–209, apr 1962. https://doi.org/10.1147/rd.62.0200.
25. H. Hernandez, D. Carvalho, B. Sanches, L. C. Severo, and W. V. Noije, "Current mode 1.2-Gbps SLVS transceiver for readout front-end ASIC," in *2017 IEEE International Symposium on Circuits and Systems (ISCAS)*, May 2017, pp. 1–4. https://doi.org/10.1109/ISCAS.2017.8050227.

Chapter 4
Verification and Testing

The design presented in the previous chapter has been extensively tested to guarantee that it functions as expected once produced. This chapter presents the methodologies and results of the verification and testing work that was completed for the first, second and third prototype. The chapter opens with a discussion on the verification work done on the digital design through the use of software simulations and modelling of the design features. Methodologies in the design process to reduce the probability of bugs appearing in the design are also discussed. The chapter continues with a description of the acquisition system that was designed for the hardware tests and verifications. The focus is then moved to the hardware device validation and testing with a primary focus on the tests that have an impact on the digital design. Other tests, e.g. analogue and ADC characterisation, have also been completed, but will not be discussed here, as they don't directly affect the digital design.

4.1 Functional Verification of the Digital Design

Functional verification is the task of checking that the system operates as intended and as specified. It is one of the most important tasks in the design of ASICs, due to the cost and turn-around time involved in their production. If an issue with the digital design is detected after the device has been produced, it can be very difficult to determine the cause of the issue, as there are usually no simple ways to observe the internal operation of the design, unless specific debugging features have been implemented for this purpose.

In contrast to analogue designs, digital designs can be almost fully verified through computer simulations as long as the design itself is fully synchronous. In an ideal synchronous design, the state of the system only changes when triggered by a clock signal and so the state can be predicted by the initial state of the system and the stimuli provided. The state of an analogue or asynchronous design is more dependent on how well the physical, manufacturing, and environmental properties are modelled.

A. Velure, *A Digital Signal Processor for Particle Detectors*, Springer Theses,
https://doi.org/10.1007/978-3-030-71559-5_4

As long as the synchronous design operates well below the speed limits of the process technology, it approaches an ideal synchronous design. The parts that are more difficult to simulate are the interfaces to any internal analogue parts and interfacing to the outside world. Mixed-signal simulations are used for this purpose; where in the analogue-parts are either modelled in a simplified manner, from only a behavioural perspective, or the full analogue schematic is modelled.

To drive a digital simulation, a testbench code is created, which encapsulates the design-part to be tested. Its purpose is to direct the overall testing effort, as well as to provide stimuli to the device under test, and to verify automatically the correctness of its response. It is essential to have tests that verify the complete operation of a design, including all corner cases, to minimize the risk of discovering issues after production. Code/test coverage tracking helps in this regard, where the tool collects information on which statements have been executed in the code, which branches have been taken, which states that have been covered or transitioned between, and which bits in a register have been toggled. Code that cannot be covered by exercising the inputs is likely to be dead code and might need to be removed. Exceptions would be, for instance, SEU hardening code like TMR, which is triggered by impinging particles.

Code coverage only verifies the correctness of what has been implemented and can be generally considered a quantitative measure of the design code, but it does not necessarily verify that all the specified functionality has been implemented or tested. The design intent from the specifications must, in addition, be formulated as testable statements that can direct the testing to verify that all features have been covered, generally referred to as functional driven verification.

To detect illegal transactions and signalling in the code, the design uses assertions, which are small pieces of simulation code inserted in the design code that reports when certain conditions occur. The benefit of including assertions is that they report both when testing a module standalone and when it is tested together with other modules as part of a bigger hierarchy. With assertions, interfaces between modules can more easily be verified when testing a bigger hierarchy, and the origin of the error is easier to pinpoint without the need to trace an error from a top-level output back to its point of origin.

The testbenches in this design use a combination of directed testing, where the specific design intent and specifications are verified, and random testing, where the input stimulus is randomized to exercise the bulk of the code coverage metrics. Testing the complete behaviour of a sub-module from the topmost level of the design hierarchy is often complicated by both the limited set of input controls available from the top level and the increased simulation time that is required to test the larger amount of code. Due to this, it is generally preferable to test the sub-module in isolation as well. However, since testing in isolation might not capture all the intricacies of the inter-module communications in the design, it is still important to try to verify as much as possible from the top level, particularly since the top-level testbench is used for verification of the post place and route code including timing verification. For the v2 design, the top-level testing was primarily done using functional directed testing, but more randomized tests where added for v3 as issues were discovered during hardware testing of v2 that were not caught during simulation, because the functional coverage was not exhaustive or high enough.

For the design of the SAMPA v2, the design was primarily tested using a clear box approach, as opposed to a black box approach, where the designer of the tests also was the designer of the circuitry to be tested. The benefit of this approach is that the needed coverage is reached faster since the designer knows which specific issues and problems to test for, but the designer may, on the other hand, have an obstructed view on other potential issues due to having pre-knowledge from the design phase.

To provide a consistent testing environment, this design uses the Bitvis Utility Library [1], which is an open source VHDL testbench infrastructure library. The library reduces the workload needed to design and analyse tests by providing a structured uniform logging mechanism. It also provides methods for checking and reporting of signals and transactions with verbosity and severity handling. The test-benches are compatible with and can be run on both Cadence Incisive Simulator [2] and Mentor Questa Sim [3].

To aid in code review and to help visualize the designed implementation, all the main modules have additionally been drawn schematically and are available in [4]. Based on the results of this process, various pieces of redundant or unnecessary circuitry were discovered and removed between v2 and v3.

The design has been functionally verified through simulation with TMR protection enabled, but verification of the correctness of the TMR implementation and its ability to correct for error has not been done in simulation due to lack of suitable tools and the complexity involved in doing the verification manually. Verification has instead been done in hardware; see Sect. 4.3.4 and [5].

4.1.1 Tool Based Analysis

To detect all problematic edge cases in a design, the amount of time needed for simulation grows rapidly as the design grows in complexity. To alleviate this, certain tools are available which runs static analysis of the design code to detect potential issues. Instead of simulating the design, the tools use techniques like data flow analysis, control flow analysis, lexical analysis, and formal analysis to find potential issues.

4.1.1.1 Static Code Analysis

The RTL code for the digital design has been written in Verilog with some enhancements from System-Verilog. Verilog is a loosely typed language, which means that any data type can be assigned to a variable of a different type without an error or warning being created. The benefit of this approach is that the code becomes less verbose and easier to write, but the drawback is that it is relatively easy to make a minor mistake that is hard to detect even with the help of simulation. There are also certain idiosyncrasies to the language that makes it prone to produce bad code [6]. As such, it is beneficial to use programs that statically analyse the code for suspicious constructs like variables that are never read, truncation of a signal when it is assigned to another variable, not taking care of overflow in adders, etc. These tools are commonly referred to as linting tools. The tool can also check that the code adheres to certain ways of coding that are less prone to produce problematic code. Some tools additionally incorporate formal-verification techniques to detect for instance unreachable states in a state machine.

Some linting behaviour is commonly built into the synthesizers, but their analysis is not as exhaustive as with stand-alone tools. Since there are variations on the different checks that different tools do, this design used Verilator [7], Mentor HDL Designer [8] and Cadence Incisive HDL Analysis and Lint (HAL) [9]. The tools have a tendency to report several false positives, for instance, if you bring a packed configuration-register vector into a module, but only use parts of the vector in that module, the remaining bits will be reported as unused. Other types of warnings that might present themselves are if there are unused outputs on a module, or if there are registers without a reset. For v3 of the SAMPA, Verilator reports 80 warnings and HAL reports 2 errors and 785 warnings. Numbers for HDL designer is not available as the tool no longer is able to parse the code for SAMPA v3 due to the introduction of more System-Verilog code in v3, which is not as well supported by it, but the numbers for an earlier analysis was on order of what HAL reports.

When the log has been parsed thoroughly once, it is later possible to do only a diff between the previous log and the current to detect any changes. Running the linting tools before new changes are committed to the repository reduces the time spent on unnecessary debugging of code. For instance, a warning will be reported if the bit-width of a register that is distributed to several modules was changed in one module, but not updated in all places where it is instantiated or used.

4.1.1.2 Formal Verification

After converting high-level RTL code to lower level gate-level code in the synthesis process, there is a need to verify that the converted code has the same functionality as the original source. This can be done through simulations, but simulations are only as thorough as the testbench itself. Logic Equivalence Checks (LECs) are instead preferred to simulations as LECs are an exhaustive analysis of all logical possibilities. A complete analysis with LEC is also faster than an exhaustive testing through simulations,

To provide redundant testing, this design used both Mentor FormalPro [10] and Cadence Conformal Equivalence Checker [11] to verify the equivalency between the RTL, the synthesized code, and the post place and route code.

4.1.2 Module Based Testbenches

Module based testbenches have been created for most of the sub-modules of the design, to more extensively verify a larger part of their functionality than what could otherwise have been done from the top level. Most of the testbenches follow the structure as shown in Fig. 4.1a where there is a main test sequencer that runs a sequence of tests to verify the design, followed by a data generator driven by the sequencer, which generates the stimuli to the Device Under Test (DUT). Running in parallel with the DUT there is a simplified behavioural model of the design, which generates what are to be the expected output signals from the DUT. A checker process verifies that the data generated from the DUT and the behavioural model matches and will otherwise report any inconsistencies. Depending on the complexity of the DUT, the data generator and the checker might be in the same process loop as the sequencer. This is particularly the case for the testbenches of the pipelined modules in the channel pipeline (pre-trigger, Digital Shaper, BC1, BC2, BC3, Zero suppression) where only a new sample value is inserted and a new baseline value verified per cycle.

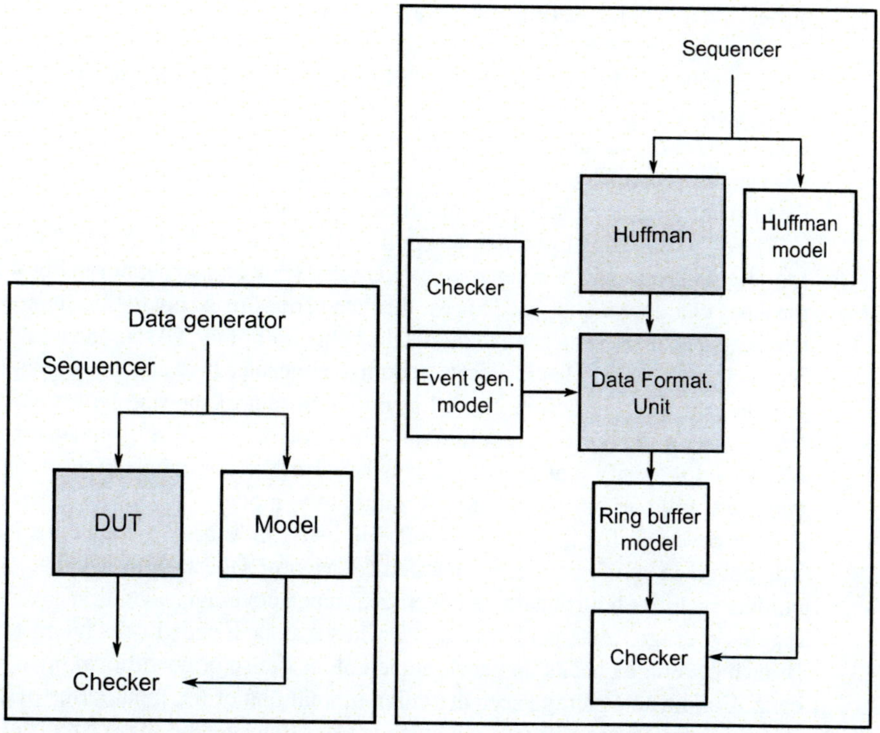

(a) *General* (b) *Huffman*

Fig. 4.1 Module based testbenches. Grey boxes are modules from the design

For interfacing to some of the DUTs, Bus Functional Models (BFMs) have been created, which abstract the multi-cycle interfacing to the module into a single call to make the testbench sequences both easier to write and read. This was done for the JTAG testbench as well as the I^{2}C testbench. The JTAG module deviates from the general testbench in Fig. 4.1a by an additional possibility to connect two JTAG modules together to verify that JTAG chaining works. The I^{2}C module deviates in that it interfaces to a generic wishbone attached memory and can be driven from both the I^{2}C BFM and a module containing the I^{2}C master from the GBT-SCA project, the same device as will be used in the experiment. A separate BFM is used to interface with the GBT-SCA master I^{2}C unit. In this way, the compatibility with the GBT-SCA can be verified before the chip is produced.

The testbenches that verify parts of the data-formatting unit and ring buffer are more complex as they need to model more of the surrounding circuitry. For the Huffman testbench shown in Fig. 4.1b, the sequencer generates sample data that is sent to the Huffman DUT and the Huffman model. The model stores the compressed data in a memory array and this data is used to compare against the output of the Huffman DUT by an independent checker process. An event generator model controlled by the sequencer generates the necessary signals to simulate events with varying length and spacing. The output of the data-formatting unit is fed into a model of the ring buffer and a checker verifies the data in the memory against the data produced by the Huffman model earlier.

The testbench shown in Fig. 4.2a takes a slightly different approach as it is based on the SystemVerilog method of designing testbenches with a separate driver, monitor, sequencer, and checker.[1] The driver generates both the stimuli and models the two DUTs. The monitor models the memory readout behaviour and provides the data to the checker, which verifies it against the model data.

The daisy chaining testbench was built in two steps. In the first step, the data from the neighbouring chip was generated by a model (link data generator in Fig. 4.2b). As the daisy chaining is supposed to pass the data verbatim, except for the suppression of sync packets, there is no need to model the behaviour of the DUTs, and a FIFO buffer is used instead. A checker is present at both the interface between the neighbour input module and the ring buffer, as well as on the output of the ring buffer. For v3, the RTL code for the serial output module, the heartbeat generator, and sync generator was added to verify that all inter-delays between the communication of the master and slave unit had been taken into account in the model.

The coverage for the tested modules is listed in Tabel 4.1 for both v2 and v3. The coverage numbers were generated by Mentor Questa Sim [3]. The coverage number includes both statement coverage, branch coverage, condition coverage, and toggle coverage. Statement coverage verifies that each code line is hit at least once. Branch coverage checks that each branch of an if/else or conditional operator is hit once. Condition coverage sees that all combinations of the input terms of a branch statement are covered. Toggle coverage verifies that each bit of each register or wire has transitioned from 0 to 1 and opposite at least once.

[1] DFU ring buffer testbench created by Raul Acosta Hernandez, University of São Paulo, Brazil.

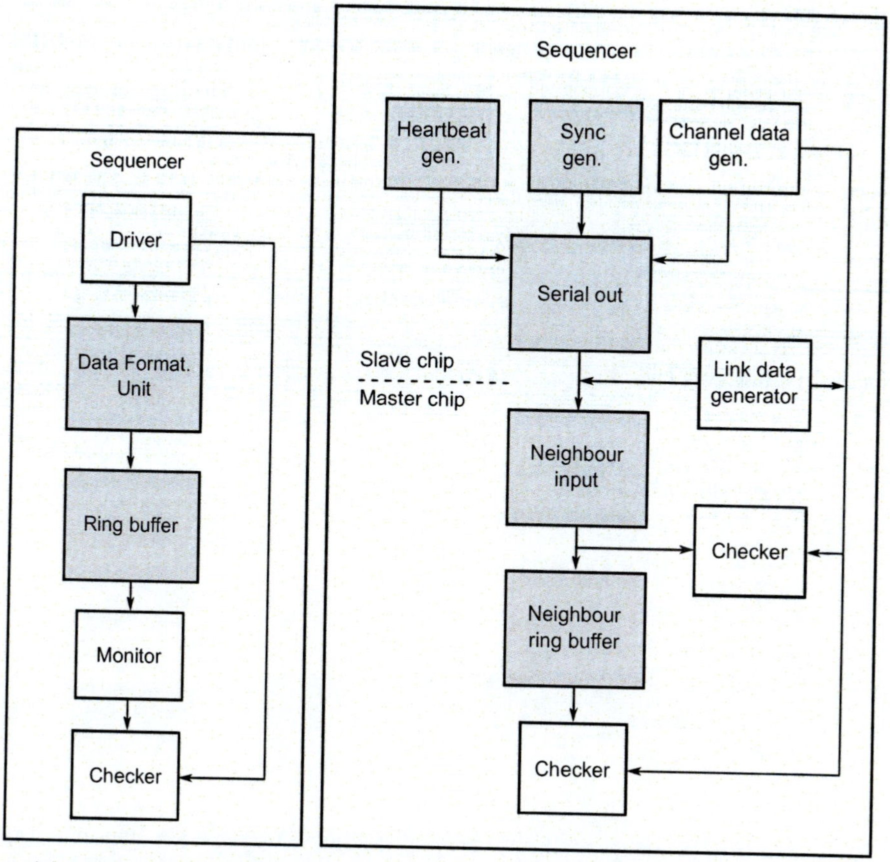

(a) *DFU and Ring buffer* (b) *Daisy chain*

Fig. 4.2 Module based testbenches. Grey boxes are modules from the design

An effort was put in to increase the coverage of the design for v3. In v2, the lower coverage was mostly caused by dead or unreachable code branches that were removed for v3. In other cases, it was found that not all bits in a register had been transitioned during the testing to cover all parts of the design. In v2, the data-formatting unit and ring buffer were only verified on the top level and so the coverage numbers are not included.

4.1.3 System Level Testing

The testbench for the top-level design is sketched in Fig. 4.3. The testbench can be set up for testing of a single SAMPA or for testing of two SAMPAs in daisy-chained mode by changing a parameter at compile time. When testing only a single

Table 4.1 Code coverage summary of module based tests

Module	Total coverage v2 (%)	Total coverage v3 (%)
Pre-trigger	98.1	100
Digital shaper	94.2	95.7
Baseline Correction 1	96.6	96.9
Baseline Correction 2	95.5	93.1
Baseline Correction 3	100	100
Zero suppression unit	98.5	100
JTAG	82.2	100
I^2C	98.8	100
Event manager	86.3	100
Huffman	94.6	100
Neighbour control	100	100
Neighbour ring buffer	99.5	99.5
Data-formatting/compressing unit	44.9 (only Huffman)	100
Ring buffer input/output	–	100
Clock generator	98.1	100
Ring oscillator	98	100
Hamming encoder	100	100
Hamming decoder	100	100

SAMPA, the boxes marked in grey are excluded from the simulation to increase the simulation speed. By changing another parameter at compile time, it is possible to run the simulations with gate level code, both with or without back annotated timing. A wrapper encapsulates the SAMPA code to model some of the analogue behaviours, like for instance the I^2C tristate driver and the SLVS enable signals.

The serial link data checkers operate independently from the test sequencer. In case of certain tests where checking is not needed, they can be disabled to avoid generating unnecessary errors. The task of each checker is to automatically synchronize to the incoming data, like a real serial receiver, do integrity checks on the packet header, verify the header and data are without any parity or Hamming errors, compare the received header and data against the next expected header and data received from the behavioural model, do a decompression test of the data to verify that the compressed data fits with the raw input data and finally do a graceful self-recovery in case unexpected errors occur, so that testing can continue from the next packet.

If the serial link checker has been marked as not being the last checker in a chain it additionally puts the output data in a buffer to be checked by the next checker in the chain. Since each serial link checker has knowledge of which chip address it receives data from, it ignores the packet that is not from the expected chip address and forwards them directly to the buffer.

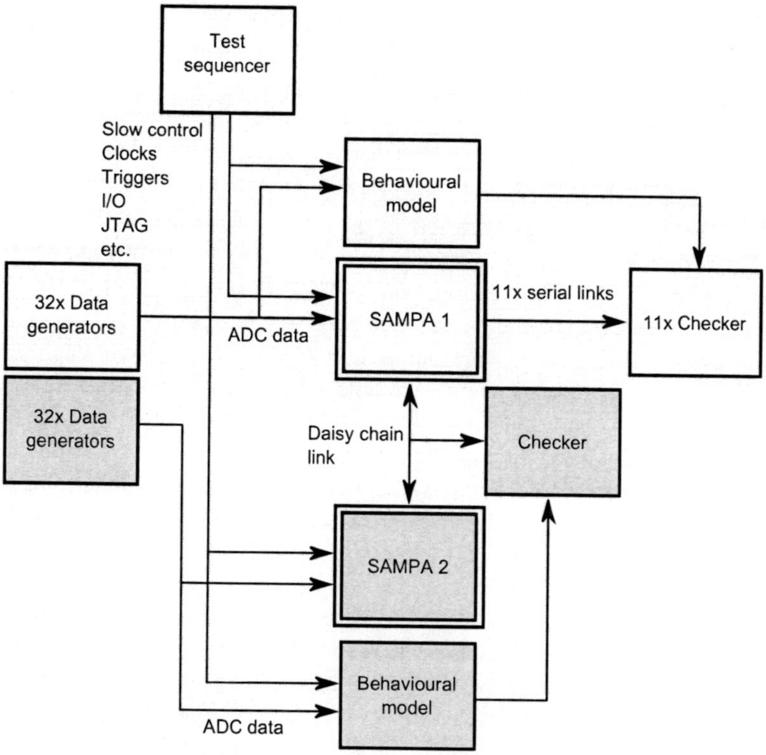

Fig. 4.3 Testbench for top-level design. Grey boxes only included when doing simulation of daisy chaining. Double box on SAMPA indicates a wrapper for emulating some of the analogue behaviour

The behavioural model for the SAMPA contains separate processes for modelling of the clock generator, the event generator, the reset generator, and the 32 channels including filtering, compression, and memory handling. Modelling of the channel pipeline is cycle-accurate and in phase with the pipeline in the SAMPA so that debugging of issues can be easily done by comparing the data in the pipeline between the model and the real code. The behavioural model tries to emulate the memories of the SAMPA as close as possible so that it takes into account overflows. The memories and its pointers are accessible from the checker through use of shared variables.

The data generators can be set up to generate different sequences of signals depending on what is to be tested. It can produce full range sawtooth waves, which aids in debugging since it makes it easier to compare the output data of the model and the real circuitry when an issue occurs and determine if something happened too early or too late and by how many cycles. It can also be set up to run with random input data, which by adjusting the zero suppression threshold of the filters can easily emulate different data occupancies for verification of buffers and serial links. Additionally, it can read plain CSV files with sample data or custom data files from for instance TPC physics runs in 2010, which has both normal data as well as what is referred to as "black events", which have data from collisions with an exceptionally high amount of particle tracks.

As the testbench has been designed modularly, it is possible to include several devices in a chain if needed. This only requires instantiating new modules and routing the intercommunicating signals, if needed. However, the MCH only required two devices in a chain for the final design [12].

The BFMs previously created for the module-based testbenches are also reused in the system level testbench. This includes the JTAG and I^2C BFMs. Additional BFMs have been created for reading and writing from both the global register, channel register, pedestal memory, and channel order register.

Configuring of the device happens through I^2C and since it only operates at maximum 1 MHZ, it takes significant time to set up configurations for different tests. To avoid this, the testbench instead forces values onto the internal signals between the I^2C module and the register module inside the SAMPA once the testbench has been confirmed to be working satisfactorily in a separate test. Forcing and spying on signals is a feature available in the 2008 version of VHDL, but the Cadence simulator has unfortunately only limited support for VHDL-2008. Both Mentor and Cadence, however, support these features through their own special function calls that are not part of the standard. To make the testbench environment uniform across simulators, a generic package per simulator has been created. It wraps each of these special function with a generic function name so that the testbench can call the functions with a generic name and during compile, only the package file for the specific simulator is included which has the version for the generic function translated to the internal one. In this way, the testbench can be simulator agnostic.

The testbench sequence is mainly focused on verifying the functionality of the design and the extensive verification is left to the module-based testbenches. The testbench verifies that all interfaces to the device functions, this includes I^2C, JTAG, memory tester, scan chain, clocks, resets, and triggers. It also verifies all data output modes, this would be the direct readout serialization, the packet based serialization, and the daisy chaining. Additionally, it verifies that all test-modes like the bypass mode, ring buffer, direct readout combinatorial, etc. operate correctly. Overall, this covers the operational functionality of close to everything that can be controlled externally.

4.1.3.1 Scan Chain Verification

To minimize the workload and to avoid errors, the scan chain insertion, see Sect. 3.3.1, was done by a tool (Cadence Encounter Test, now Modus [13]) instead of manual insertion. In addition to the scan chain insertion, the tool also creates vectors for testing the design and creates a Verilog testbench that runs the complete vector test on the digital code. To verify that the generated vectors do not produce any false errors, the testbench was run on the gate level code with timing. To additionally verify that the fault-finding algorithm correctly detects an error, a wire was deleted from the design and the testbench was rerun. An error was produced as expected and the error was indicated to be at the point where the error had been introduced.[2]

4.1.3.2 Clocking and Clock-Domain Crossing Verification

Signals crossing between two asynchronous clock domains might experience metastability, which could cause incorrect data to be generated [14]. Since the SAMPA design employs four clock domains where some of the clocks might be asynchronous if multiple clocks are externally sourced, it is both important to code the design to avoid any metastability and to do verification to confirm that all crossings have been checked. The design uses two-register synchronizers for all signals passing between two clock domains [14]. Pulsed signals passing from a fast clock domain to a slow one are converted to signals that toggle on a rising edge before being passed to the slow domain, this still requires pulses to be separated by two or more cycles on the receiving end, but does not require the signal to stay high for multiple cycles. Multi-bit signals from counters without control signal are Gray-encoded before being passed between clock domains to avoid capturing the wrong value if the signal is captured in the receiving domain while some of the bits are still switching over from one value to another. When a control signal is present in addition to the multi-bit signals, only the control signal is synchronized and the data is expected to be valid for enough cycles to sample valid data.

To avoid that signal are inadvertently being used in another clock domain without first being synchronized, a coding convention has been employed. For modules where multiple clock domains are present, the variable names of all signals have been appended with a postfix indicating the clock domain that the signal belongs to, e.g. signal_clkadc, this avoids that errors are introduced while coding, it also aids in spotting these errors while reviewing the code.

To verify correct operation of the design with asynchronous clocks, the top-level testbench was run with the device set to operate with all external clocks and the input clocks set to frequencies that were not a factor of each other. The cell library was modified so that when a metastability is detected in a flip-flop it will randomly produce a 1 or a 0 on its output instead of outputting 'X' which will propagate through the design and prevent further testing. The regular set of tests were run without issues.

Testbench verification will only catch a subset of the possible clock domain crossing issues so to additionally verify that all crossings have been covered, Mentor Questa CDC [15] has been employed. This tool uses structural analysis of the code to detect clock domains and synchronizers and can report on any potential failure modes.

The Clock Domain Crossing (CDC) testing uncovered no specific issues, proving that the employed design methodology prevented these issues.

[2]Scan chain insertion and testing done by Bruno Sanches and Dionisio Carvalho, University of São Paulo, Brazil.

Since the ADC requires clocks with a low amount of jitter, the clock generator was simulated on the layout level in circuit simulation and was found to have less than 50 ps of added jitter on the ADC sampling clock.

4.1.3.3 Mixed-Signal Verification

Mixed-signal verification[3] involves connecting the analogue circuitry together with the digital to verify that communication between the analogue and digital works as expected. Since the chip has been created with an analogue-on-top flow, i.e. the fully placed-and-routed digital design is instantiated as a block on the top-level analogue layout, there is a need to verify that the correct signals have been connected to the correct ports on the digital design.

To speed up the debugging, the Analogue Mixed-Signal (AMS) verification is usually done in three stages. Firstly, a behavioural model of the analogue components is created in either Verilog-AMS or VHDL-AMS, which can then be simulated together with the digital code at higher simulation speeds. This verifies primarily the interaction between the various parts.

Since the analogue behavioural model is only a simplification, there is also a need to do verification with the schematic netlist of the analogue to check that their behaviour is the same. These tests need to be kept short due to the increased time needed to run the analogue netlist simulations. Here a few samples have been tested through the direct serialization mode, through the normal packet based mode and the daisy-chained mode.

The tests also need to be run with the digital gate level code with back annotated timing, together with the layout version of the analogue, to verify that the timing between the ADC and the digital is in order and that the power-up sequencing is working as it should.

The mixed signal verification helped to uncover issues with signals between the ADC and digital having inverse direction, start-up issues related to the clock gating employed in the direct serialization mode and timing issues between the ADC and digital inputs.

4.1.4 Test Coverage Improvements

As explained in Sect. 4.3.4, an irradiation test was performed on the design to verify its tolerance to SEUs, but since it is difficult to do an exhaustive test of the failure rate of the design in all possible configurations during a beam test, a simulation-based effort to evaluate the effectiveness of the SEU protection could have been of benefit. This is commonly done by modifying the test environment to support fault injection

[3] AMS simulations conducted by Tiago Weber for v2 and Heiner Alarcon for v3, University of São Paulo, Brazil.

while running the normal verification testbench. Due to time constraints and lack of automatic tools to do the verification, this was not completed, though the irradiation test did not directly uncover any issues that could have been solved by fault injection.

In addition to being able to predict the failure rate of the design, fault injection can also be utilized for improving mitigation strategies by highlighting areas of the design that are more sensitive to failures than others. As the TMR technique that has been employed for SEU protection requires significant area, the test could have also helped determine areas that could run without TMR or that could employ other methods of protection, reducing the needed chip area of the digital part.

The testing of the top-level digital design was, due to time constraints, primarily done by directed functional testing of the functionality that the TPC and MCH required and so certain aspects of the design that are outside the scope of their requirements will not have been covered. The remaining effort lies in implementing models of the yet to be tested circuitry so that more of the design can be tested with randomized testing.

4.2 Hardware Verification of the Digital Design

To verify the various performance specification of the SAMPA, a continuous Data Acquisition (DAQ) system has been developed. The FPGA-based DAQ system is specially designed for the qualification of the SAMPA and is designed around an Altera SocKit [16] evaluation board containing an Altera Cyclone V System-on-Chip FPGA [17], which has a built-in dual-core ARM[4] Cortex-A9 microprocessor unit. The SAMPA is mounted on a separate mezzanine board connected to the development board through a high-speed connector. Control of the SAMPA and the FPGA board is handled through a Universal Asynchronous Receiver/Transmitter (UART) connection from the controlling computer. Data packets from the SAMPA are acquired by the FPGA and transmitted verbatim to the readout computer via Gigabit Ethernet. A block diagram of the system design is shown in Fig. 4.4.

The intention of the DAQ system is to provide an easily customizable, stand-alone platform that is both compact and easily deployable to multiple testing groups for analogue behaviour qualification, digital verification, testing of the radiation tolerance and mass production testing. For the radiation tolerance testing, the intention is to verify the device performance in regards to SEUs, SELs, and functional interrupts. For the mass production, the goal is to filter out bad devices before mounting them on front-end cards. DAQ systems and testbenches for the intention of qualifying of ASICs are commonly designed for a single or a limited set of tests. The system presented here is on the other hand designed as a common versatile platform and framework to interface with the SAMPA ASIC for a multitude of tests.

A custom DAQ solution was chosen due to a number of reasons. The SAMPA uses SLVS differential links [18] for its high-speed serial communication links, since this

[4]Advanced RISC Machine.

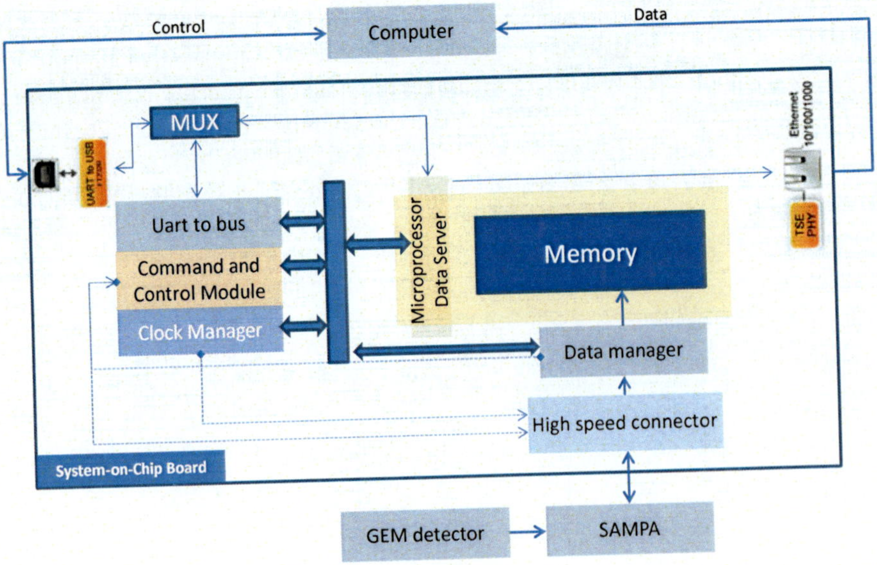

Fig. 4.4 Schematic overview of the FPGA design showing various system blocks

is also the type of link used by the Gigabit Transceiver ASIC (GBTx). The GBTx ASICs were also not yet available at the time of the production of the SAMPA v1 and Altera's FPGAs were at the time the only FPGAs that supported SLVS receivers natively. The board has furthermore support and commercial availability for extension-cables for connecting between the development board and a mezzanine board, which is required for the irradiation testing. Lastly, the development board is also available at a low cost, which lowers the entry point for people to work with the device. Development of a fully custom FPGA board was ruled out due to the complexity and time involved in doing the design. A picture of the SAMPA mounted on the mezzanine board and the accompanying FPGA DAQ board is shown in Fig. 4.5.

The system, as designed, supports both the conventional method of data acquisition where a limited amount of samples are collected in a triggered fashion before being readout, as well as continuous acquisition by using direct memory access to an embedded Linux system with a Gigabit Ethernet connection [19, 20]. To transmit the data from the DAQ to the controlling computer, the Gigabit Ethernet is used since it is available on most computers and makes the system more versatile. Optionally there are two Small Form-factor Pluggable (SFP) connectors mounted on the mezzanine board, to enable forwarding of data to a computer that can accept this. A solution using PCIe or optical fibres [21] might have provided higher readout speeds, but would require installing the card in a desktop computer or connecting it through an optical-to-electrical interface board. Crate based DAQ solutions using VME/ATCA/μTCA/PXI are also commonly used for their high-speed capability [22], but they are more expensive and considered less versatile.

Fig. 4.5 Picture of the test board for v2 of the SAMPA (right) connected to the FPGA DAQ system (left)

The embedded Linux system has full access to the stream of incoming data as well as the possibility to take over the control of all pins on the SAMPA, providing a lower threshold for non-FPGA developers to develop test and verification applications directly on the embedded system without the need to use the UART connection.

The data samples are analysed in real-time using the CERN ROOT [23] data analysis framework to monitor the data quality. To control, configure, and monitor the SAMPA and the FPGA board, a software package with a graphical user interface has been developed.

As it is possible to take over the control of the pins to the SAMPA from the control program on the computer or through the embedded Linux systems, custom tests for the SAMPA can be developed by non-FPGA designers through bitbanging the pins. This design methodology provides low development overhead, but the penalty is reduced performance. For tests of the second prototype, a JTAG boundary scan tester for checking connections between the FPGA and the SAMPA as well as an Automatic Test Pattern Generation (ATPG) for testing of manufacturing defects using the SAMPA designs scan chains was developed utilizing this method.

4.2.1 FPGA Firmware Design

The firmware design is divided into three main parts, a Command and Control unit, a Data Manager unit and a Data Server. It additionally contains a reconfigurable PLL that enables on-the-fly reconfiguration of the clocks used by the SAMPA to support testing of the different clocking options. The main modules are all connected to a common Avalon bus [24] along with the microprocessor. The bus can be accessed directly by the control program on the computer through a custom UART to Avalon bus bridge. A switch controlled multiplexer facilitates switching between having a serial connection (UART) to the bus system or to one of the Linux serial terminals, so the computer can access the Linux system for debugging purposes.

4.2.1.1 Command and Control

The Command and Control unit acts as the bridge between the computer and the SAMPA for slow control handling and pin control. It also operates as the main control unit for the surrounding sub-modules. This includes a master I^2C unit for the slow control communication, a module for reset generation and event trigger handling, and other modules for control and running of specialized tests on the SAMPA, such as a tester for the memory Built-In Self-Test and triangle/saw/sine wave generators for data path testing on the v1 SAMPA.

4.2.1.2 Data Manager

The data readout from the SAMPA requires a high bandwidth readout system. The first version of the SAMPA chip can operate three serial links running at only 160 Mbps, but the final version can operate with 11 serial data links running at 320 Mbps each, adding up to 3.52 Gbps total.

The first task of the Data Manager is to synchronize and de-serialize the data from each link into their native 10-bit word segmentation, verify the parity of the header, determine the length of the payload that follows and filter out any sync/filler packages so that they are not sent on. The valid de-serialized packets are further packed into 64-bit words to make processing on a 32/64 bit system more efficient. The packets are then further aggregated from the 11 de-serializers into four memory-writer modules via individual buffers. Each memory-writer module is connected to a separate high-speed 64-bit 100 MHz bus that interfaces to a 400 MHz DDR memory shared with the microprocessor. The FPGA fabric can thus support writing of the full 3.52 Gbps to memory. The complete system readout speed is primarily limited by the throughput of the microprocessor system.

For the second prototype of the SAMPA, the mezzanine cards have been fitted with two SFP connectors that are connected to high-speed transceivers on the FPGA. By utilizing two DAQ setups, it would be possible to offload half of the received data to another DAQ system to increase the collective bandwidth, or they could be interfaced directly to the CRU once it is available. This option was eventually not utilized, as reading out large amounts of raw data over a long period of time was not needed.

The system can be set up to acquire data continuously or to acquire a configurable number of consecutive packets per serial data link. In case an overflow in the buffers or the memory is detected, the system will drop the payload of the packet and set an overflow bit in the header, indicating to the data analysis program that the amount of data should be reduced. This is done by utilizing one of the extra bits available when 60 bits are packed into 64-bit words before the buffers. When a buffer is almost full, the module will set this bit to one and wait until the next packet. When the truncated packet arrives at the memory-writer module, this bit will be detected in the stream and the first 64-bit word in the packet will be updated with another bit that indicates to the later systems that only the first 64-bit word (the header) is valid and that the payload size as indicated in the header is invalid. The program on the Linux system that forwards packets to the computer would then see this bit as being set and only send the header instead of the full packet.

As the incoming serial data arrives at a high speed, there is a possibility that the data will be sampled at the changeover point, generating invalid data. To prevent this, the system uses the DDR registers available in the input pads of the FPGA to provide the value at both the rising and falling edge of the system clock. An auto adjustment function in the control program on the computer selects which input to use for the de-serializer based on the ability of the de-serializer to synchronize to the incoming data stream. In the packet-based mode, the serial links are operating independently so there is no need to have the incoming data in phase across the links, but in the direct readout mode, this is necessary. To compensate for phase differences of plus/minus a clock cycle, an extra register can be selectively included to the input path of each serial link. The FPGA that was used does not have run-time-configurable delay compensation in the input/output cells and implementing delays with fabric cells was not found to be reliable. Oversampling was also not an option as the device is not able to operate at much higher speeds than the 320 MHz that is the maximum speed of the serial link.

4.2.1.3 Data Server

The data server is a program running on the embedded Linux system of the microprocessor. It is in charge of handling the data transmission and TCP/IP connection control with the remote analysis program on the computer. As the data from the SAMPA is packet based, with a header and payload, it can be transmitted verbatim to the controlling computer with Transmission Control Protocol (TCP), minimizing overhead. The data throughput of the DAQ system is primarily limited by the throughput of the underlying embedded Linux system and the Gigabit Ethernet. The throughput using TCP was tested with the iPerf3 tool [25] to be 670 Mbps between the Linux system and a computer, which is lower than expected for a 1 Gbps connection. User Datagram Protocol (UDP) is generally faster than TCP, as it does not automatically re-transmit lost packets, but testing with UDP showed only a slight increase in speed to 690 Mbps with a 1% packet loss, indicating that the issue lies in the microprocessor to Ethernet interface. The extra work of manually implementing packet re-transmission to work with UDP did not seem worth it for the small speed gain. In contrast, the transfer speed from the computer to the DAQ was measured to be 950 Mbps with 43% packet loss for UDP and 716 Mbps for TCP.

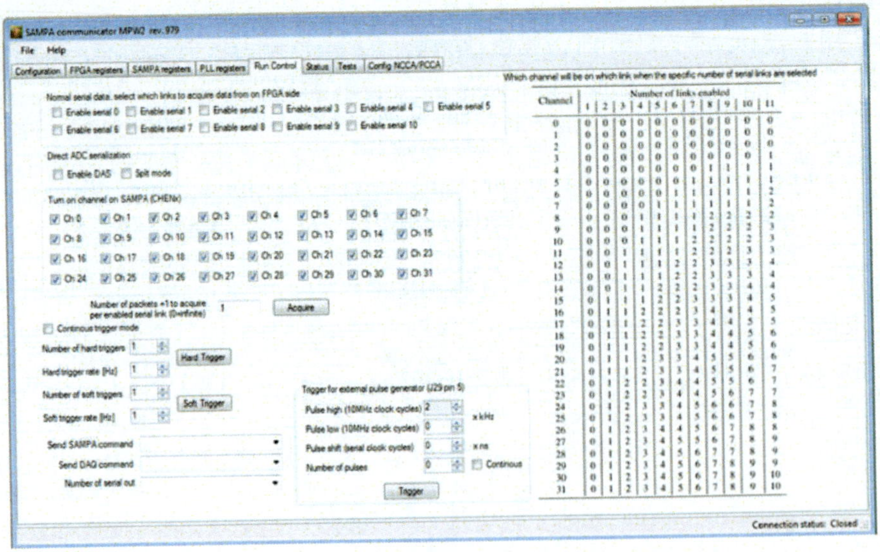

Fig. 4.6 Picture of the run-control part of the SAMPA communicator graphical user interface used for testing the SAMPA

Three ADCs running at 10 MHz can only produce 300 Mbps of continuous data, so the bandwidth is therefore not a limitation for testing the first prototype. On the second prototype with 32 channels, this problem is overcome by a form of gating. The user configures the DAQ to forward only a limited amount of consecutive packets. By keeping this number below about 65 000 packets, the system buffers will not overflow, and the system can be tested in continuous mode.

4.2.2 *Data Acquisition and Analysis*

The acquisition and control programs are run from a Linux or Windows computer and interfaces with the FPGA board through UART and Ethernet. Splitting of the programs into two separate entities is done for ease of design.

4.2.2.1 SAMPA Communicator

The SAMPA Communicator is a graphical user environment, written in C#, to handle control and configuration of the DAQ and the SAMPA. A picture of the interface is shown in Fig. 4.6. It is designed for ease of use by non-technical users, to aid in setting up and running tests on the SAMPA without the need to know details of the operation of the DAQ system. Furthermore, the program simplifies register access and provides among other things data flow handling, online status information and access to changing the frequency of the clocks supplied to the SAMPA as well as custom test and control features for radiation testing, digital verification and mass testing.

4.2.2.2 SAMPA Analyser

Data handling on the computer is taken care of by a graphical ROOT program. The program sets up the Ethernet connection with the FPGA board and receives data packets. The integrity of the packets is verified through the parity and Hamming codes present in the packet headers. Heuristic methods are additionally employed to detect anomalies in the packet coding. By decoding the run length encoding of the compression and checking that uncompressed data is not larger than the maximum time window length of 1024, the validity of the compression can also be confirmed.

The sampled signals are plotted and displayed per channel, the raw data packets are additionally written to a ROOT file for further analysis off-line. When running in continuous acquisition mode, the plots can be set to update only when the data passes a threshold value, similar to the triggering function on an oscilloscope. For each channel, both the waveform of the last acquired time window, a histogram of all samples and a histogram of the maximum value per time window is presented. The full histogram can be used for determining the noise level, whereas the max value histogram can be used to determine the amplitude of an injected pulse. Fitting of a 4th order Gaussian shape can be applied to pulses in the waveform display to get the shaping time and amplitude of the received pulse.

4.2.3 Design for Test Features

For design simplicity, both the JTAG and scan-chain tests are implemented in software and the control signals are bit-banged through interfacing with the Command and Control module on the FPGA. The JTAG tester is implemented directly in the SAMPA Communicator program and runs walking one and walking zero patterns across the input/output pins of the SAMPA to detect stuck pins, inverted pins, or shorts. The scan chain tester is, on the other hand, implemented as a program running on the embedded Linux system. This is done to save on testing time by avoiding to pass data through the slow UART-to-bus interface. Running directly on the Linux system provides a maximum speed close to 1 MHz compared to a few kHz when running from the computer. The scan chain program is a port of the Verilog testbench, discussed in Sect. 4.1.3.1, into C and uses the same test-vector files as used by the testbench [26].[5] The tester for the built-in memory tester could also have been implemented in software for the manufacturing defect tests, as the SAMPA has a go/no-go output to indicate if there is an error. However, to be able to reuse the tester

[5]Testbench conversion done by Bruno Sanches, University of São Paulo, Brazil.

for irradiation tests, were there is a need to count the number of errors encountered, this was implemented in the FPGA firmware, since the pulses generated by the tester are faster than what can be detected from software and therefore needs to be sampled by the FPGA.

4.2.4 DAQ Verification and FPGA Prototyping

Testing of the complete readout chain in a simulation is difficult due to the complex interaction between the SAMPA, the FPGA firmware, the ARM processor in the FPGA, the Linux system and the control and analysis software on the computer.

The FPGA firmware can by itself be verified through simulation by reusing the top-level testbench used for verification of the SAMPA and combining it with testbench functions, which verifies that the FPGA firmware can de-serialize and aggregated correctly into memory.

To verify the full readout chain, the Verilog code for the digital part of the SAMPA was made synthesizable for the Altera FPGA and synthesized into the top level of the FPGA code. While running the system, selecting between communicating with the external SAMPA and the synthesized internal SAMPA can be done through a switch. For the first prototype, the full SAMPA was synthesized for use in the FPGA, while for the second prototype a reduced version of the design with only eight channels was used. The SAMPA design utilizes several large multiplexer structures, which prohibits the full design to fit on the utilized FPGA.

The only modifications necessary to make the SAMPA code synthesizable on the FPGA was to modify the TSMC cell library so that custom cells were replaced by normal cells. Additionally, the FPGA synthesizer was told to use memories based on registers instead of the original SRAM IP memory. For efficiency, the synthesizer automatically infers block memory when large register-based memories are found. The option to use register-based memory was already defined in the original code as it is utilized for running faster simulations during debugging of testbenches. Features that rely on the specific delay of cells were not taken into consideration during the conversion i.e. neighbour input delay compensation, ADC clock delay compensation and the ring oscillator. If these features were required, they could have been adapted to use the vendor provided delay cells of the FPGA.

To be able to verify that the connections between the DAQ board and the mezzanine board containing the SAMPA were valid before the SAMPA and mezzanine boards were produced, the synthesized SAMPA digital code was implemented on a separate FPGA board and the two boards were interconnected through a cable. This allowed operation of the complete DAQ system readout chain as if all components were available.

An emulator of the DAQ functionality, that replicates the packets coming from the DAQ board, was made to aid in developing and verifying the functionality of the analysis program. The program can be run locally on the computer or directly on the Linux system of the DAQ board to verify the connection between the board and the analysis program. The emulator is primarily useful to verify the correct decompression of the various types of compression, the correct handling of bad or incomplete packets, verifying the correct decoding of SAMPA headers and checking that the analysis program can support the necessary bandwidth.

4.3 Validation and Testing

Three prototype production runs have been completed of the SAMPA design. The first prototype run contained three test chips. The first of which was a three-channel CSA + shaper, the second contained a single ADC and an SLVS driver/receiver pair, and the third was a reduced version of the complete design with only three full front-end channels with ADCs and a simplified digital section.

The second prototype run contained the complete 32-channel design in addition to three test chips. Since the full design does not have provision for either injecting or reading signals along the analogue processing chain, it becomes difficult to debug or verify the performance of the individual parts by themselves. For instance, the ADC does not have enough resolution to determine the analogue noise level to a satisfactory degree at low gains. The sampling speed is also too low to investigate the pulse shape of the analogue signal. To determine the Effective Number Of Bits (ENOB) of an ADC the common technique is to inject a full range sine wave into the input of the ADC. This is not possible in this context since the input signal passes through pulse shaping elements. An analogue-only test chip, which only contained the 32 CSA + shaper, has therefore been added to verify the analogue performance. The full 32 channels are included to investigate any issues on power distribution and crosstalk. Additionally, a chip containing a stand-alone ADC block is included to verify the ADC performance and another chip for the SLVS driver/receiver pair. The full chip was mounted in a 15 × 15 mm 372-ball BGA package, identical to the final package to be used for the production run. This chip was also available as a wire-bonded version. The other test chips were all wire-bonded.

More information on the test of the first prototype can be found in [27] for the analogue and ADC, and in [28] for the three channel simplified design and in [29] for the second prototype.

The test environment described in the previous section was used for verification of both the ADC and digital version of the first prototype and the complete design for the second and third prototype.

4.3.1 Test Results for SAMPA V1

Due to a design issue that caused inconsistencies in the data received by the digital section from the ADC, the analogue performance of the first full-chip prototype was not tested extensively. Some analysis is available in [28], but suitable results were not achieved unless fitting was used to get the amplitude and shaping time of the received pulses.

Originally, the issue was thought to be due to metastability in the sample value when it was transferred between the ADC and the digital section since the two sections had separate input clocks. But, by supplying a delayed clock to the ADC, that was in sync with the digital clock, it could be seen that this was not the cause as the signal quality could not be improved. Further tests on the separate ADC prototype showed that the SAR state machine was missing an output register so data was only valid for a few nanoseconds after the end of a conversion. In the design of the second prototype, the missing registers were added and the ADC was supplied with both the sampling clock and the SAR state machine clock from the digital part to avoid any metastability issues. AMS simulations were additionally completed to avoid any other possible analogue-digital issues for the next prototype.

To avoid that a possible error in the digital design would prevent the testing of the performance of the combined analogue, ADC and digital design, some external test pins were included, as shown in Fig. 4.7. Separate bypass signals were added for each filter block, in case the filter failed to work or in case the slow control interface did not operate properly, which would prevent the filters from being configured. A parallel output was also added so that the signals at the output of each filter block for one of the channels could be analysed. The parallel output could also be used to output the raw data from each of the ADCs in case some parts of the memory or serialization block failed to work. As the data on the parallel output was registered on the digital side before exiting, it was not possible to observe the ADC signals directly as on the separate ADC prototype. A parallel input data port was also present to feed the digital data chain directly in case the ADC would not operate properly.

Through the use of the parallel test input, a test signal could be applied and verified on the serial output. This proved the proper operation of the continuous triggering mode and the digital signal processing chain together with the ring buffer and serializing. The slow control, which was not I^{2}C at the time, but a custom serial protocol, was proven to work correctly by reading and writing all available registers. The filters were not extensively tested, as they were not significantly altered from the S-ALTRO implementation.

4.3.2 *Test Results for SAMPA V2*

The following section presents the results for the second prototype that contained the full-chip design, received summer 2016.

4.3.2.1 Manufacturing Defects

For the first batch of 142 tested devices, it was found that 10% of the devices produced an error during scan chain testing and no devices had JTAG errors. However, the

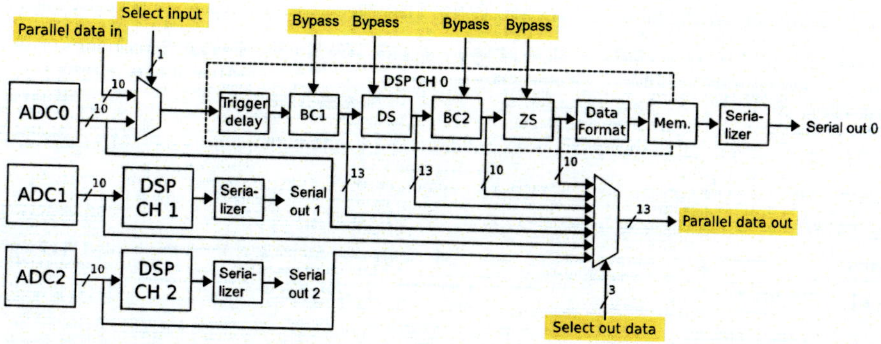

Fig. 4.7 Overview of the processing chain of the first prototype of the SAMPA digital section including debug facilities (yellow)

checker for the built-in memory test in the DAQ system FPGA firmware had a bug at the time of testing and therefore did not report any errors. Additionally, 4% of the devices showed non-working analogue channels, 3% had abnormally high power consumption or shorts, and 39% had an issue with the analogue bias reference where the voltage was out of specification. The voltage bias issues can be remedied by using an external circuitry and is therefore of less concern. The total yield for the first batch is 44% if all issues are included, or 81% if the bias reference problems is not considered. Only the value of the baseline was used for determining if a channel was broken when testing the first batch, so more broken channels might possibly have been detected if the channels had been tested with an input signal.

For the second batch, 748 devices were tested.[6] 6% of the devices produced an error during scan chain testing, 0.3% had a pin error from the JTAG testing, 6% gave an error from the memory testing and 0.9% gave both a scan chain error as well as a memory error. Additionally, 10% of the devices had analogue channel issues, 1.3% had abnormally high power consumption or shorts, and 37% had an issue with the analogue bias reference where the voltage was out of specification. Additionally, there were 2.5% with unspecified issues, which could not be attributed to either of the above issues. Overall, the total yield for the second batch is 34% if all issues are included, or 72% if the devices with only bias reference problems are counted as good devices as well.

Since the on-chip memory occupies almost half the area in the device, it is natural that there could be about the same amount of errors in the memory as in the logic itself, as seen in the testing of the second batch. The percentage of devices with both errors is also on par with what would be expected. If this reasoning is applied to the yield for the first batch, the yield for both batches would be the same, which indicates that the testing is consistent and that the scan chain and memory tester are both functioning correctly.

[6]Test of second batch was performed by Anders Oskarsson and David Silvermyr, Lund University, Sweden.

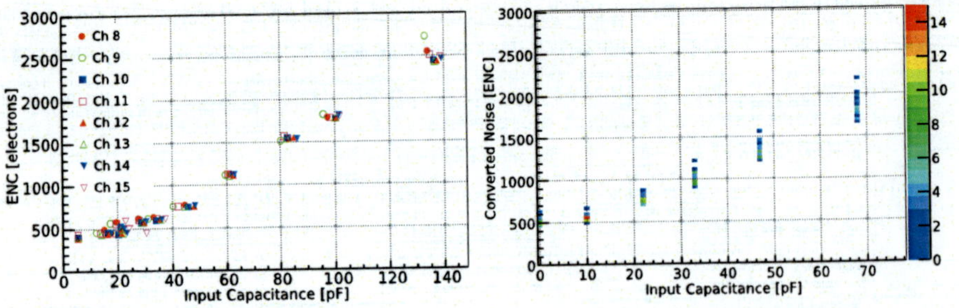

(a) *SAMPA v2 analogue only noise* (b) *SAMPA v2 full chip noise*

Fig. 4.8 Noise as a function of load capacitance at 20 mV/fC gain for the analogue-only 32 ch front-end chip (left) and the full chip including ADC and DSP (right) [29]

4.3.2.2 Noise Tests

An issue with the ALTRO and S-ALTRO chips is that the analogue section is influenced by the noise created by operating the digital section and the parallel bus readout. To combat the first issue for the design of the SAMPA, an emphasis has been put on shielding and separation between the analogue and digital sections on the layout level, as well as having separate power domains. By using low swing differential serial links instead of a single-ended parallel bus, the effect of the readout should also be reduced. Finally, an option to delay the sampling clock for the ADC in relation to the operational clock of the digital part is added to observe if this had the same positive effect as on the S-ALTRO.

In Fig. 4.8, the noise for an input capacitance of 20 pF and a gain of 20 mV/fC is compared between a 32-channel analogue-only test chip and the full chip. At the TPC capacitance of 18.5 pF, the noise is increased by about 200 e_{rms} for the full chip. The test chip was wire-bonded directly to the circuit board while the full chip was in a BGA package. As the package will add some extra capacitance to the inputs, this will account for some of the noise increase. Additionally, the ADC, which is situated closer to the analogue section than the digital, will also account for some of the noise. As the noise increase is only about 20% and the total noise is below the requirement for the TPC, it is not of great concern. A lowering of the digital noise would likely not create a significant improvement compared to the effort.

The device contains a register to configure the delay between the analogue sampling clock and the digital operational clock. The digital circuitry consumes extra current in the transitioning phase of the clock, which could spread as switching noise to the analogue section. By moving the sampling point to when the digital section is not transitioning, the hypothesis is that the sampled value should be less affected. In Fig. 4.9 the noise at different configuration setting of the delay chain is plotted,[7] but no significant change in the noise level can be seen from varying the delay between the clocks. Contributing factors are likely the distance between the digital and the

[7]Test of clock delay was performed by Anders Oskarsson, Lund University, Sweden.

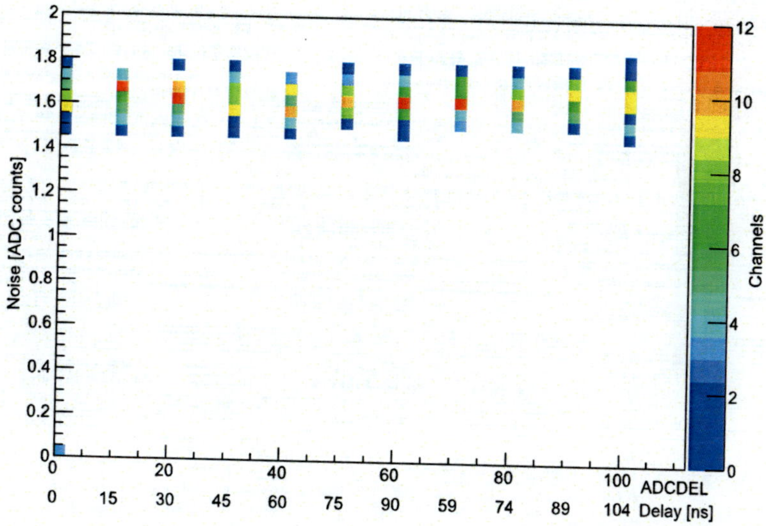

Fig. 4.9 Noise versus delay of ADC clock contra digital clock at 30 mV/fC. ADCDEL is the ADC delay configuration register setting. Each bit corresponds to about 1.5 ns delay. For values above 63 the clock is first inverted before the delay is added as if from 0

analogue, and the fact that the power-consuming parts, primarily the channel data-memories, are situated in the middle of the digital section and are as such even further from the analogue section. In addition, the fact that the digital circuitry operates on a separate power domain from the analogue and ADC helps in this regard.

In the direct readout mode, discussed in Sect. 3.2.6.3, the primary portion of the digital design is clock gated and therefore will only produce a minimal amount of switching noise. Since the output serial links are still toggling to send data, the noise generated from the links will generally be the same, if not more since the direct readout mode uses all of the serial ports. Figure 4.10 shows the noise per channel when in the normal mode with the DSP on, the direct readout mode with the DSP off and in the direct readout mode when the transmission is split into serial link groups of five. No significant change in the noise is visible between having the DSP on or off.

4.3.2.3 Other Tests

During testing of the v2 batch, the built-in ring oscillator test, that automatically calculates the ring oscillator frequency based on the sampling frequency, was run for each device. Figure 4.11 shows the histogram of the measured values. The frequency was intended to run at a nominal of 160 MHz based on the delay of the inverter cells used, but the design was not optimized to account for the wire delays when the gate-level code was available and so the nominal value is lower than originally designed. The circuitry for determining the ring oscillator frequency can, however,

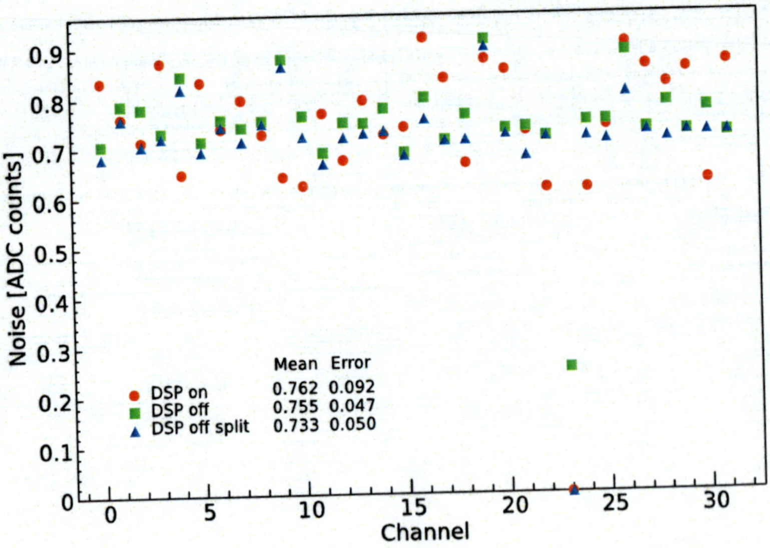

Fig. 4.10 Noise on single device for normal DSP operation, direct readout in normal mode and direct readout in split mode where the main DSP functionality is shut off. Channel 23 is dead on the board that was tested

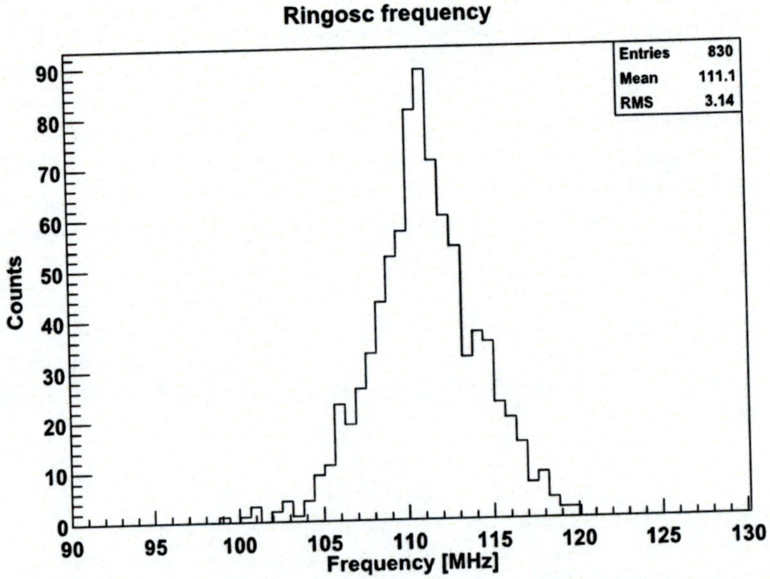

Fig. 4.11 Measured frequency of ring oscillator in v2

Table 4.2 MCH production test statistics

Test station	1	2
Devices tested	21 586	22 406
Devices passed	15 720	17 147
Devices failed	5872	5264
Fail category		
Analogue current	913 (4.23%)	721 (3.22%)
Link synchronization	301 (1.39%)	288 (1.29%)
Analogue pedestal	2186 (10.13%)	2119 (9.46%)
RMS noise	2819 (13.06%)	2375 (10.60%)
Gain	2336 (10.82%)	2045 (9.13%)
Rise time	1973 (9.14%)	1654 (7.38%)
Crosstalk	221 (1.02%)	146 (0.65%)
Ring oscillator	246 (1.14%)	291 (1.30%)
I^2C	216 (1.00%)	275 (1.23%)
JTAG	473 (2.19%)	364 (1.62%)
Memory test	1245 (5.77%)	1403 (6.26%)
Scanchain	1551 (7.19%)	1269 (5.66%)
Bit errors	546 (2.53%)	505 (2.25%)

operate with a ring oscillator frequency down to 80 MHz, and so is not affected, since the spread in values barely goes below 100 MHz.

4.3.3 *Test Results for Final SAMPA Production*

Table 4.2 presents the current results from the two test stations for the ongoing production testing of the SAMPA for the MCH as of winter 2019.[8] About 40 000 devices have been tested, with a current yield of 74.7 %. About 6.5 % of the devices have been scheduled for a retest due to inconsistent results. Previously, about 13 000 devices were tested for the TPC, but error statistics are not available for that batch. However, the experience showed that retesting devices that failed testing could bring the yield up by about 7 %-points. Primarily analogue based tests might pass on second testing, while digital test would not. The secondary testing thus might bring the final yield close to 80 %. Comparing the results with those for the testing of v2, we see they are close to the same considering the lower statistics for the v2 testing.

[8]Test of second batch was performed by Anders Oskarsson and David Silvermyr, Lund University, Sweden.

4.3.4 Irradiation Tests

Irradiation tests have been carried out to characterize the SEE behaviour of both the process, the IP memories, and the SAMPA design and to verify if the design can withstand the ALICE TPC and MCH operational environment and behave normally. Complete results from these tests can be found in [5, 30], only a brief summary will be given here. Initial characterization tests of the process and memories for SEUs showed a cross section on par with what is expected for a 130 nm process [5]. The full device also operated predictably and data could be acquired. The testing showed frequent abnormally high current increases of the device, which would disappear once the power was cycled. This behaviour is an indicator of SEL in the design. Irradiation tests had previously been performed on the first prototype of the design where no SELs were observed [30]. The primary change between the first and the second prototype is the inclusion of the memory blocks. Even though the design contains many memories of different sizes, they are all based either on an IP of a single port version or a dual port version, which both have different layouts. For the second irradiation test, the beam was collimated so that it points to either the dual port memory or the single port memory. Results showed a significant higher rate of current increases when irradiating the single port memory. So for the third version of the design, the single port memory was replaced by a dual port memory with one disabled port. Following irradiation-tests indicated that this was a success [5].

The current channel baseline value, as calculated by the BC2 and BC3 filter, is available from the configuration registers of the SAMPA. From monitoring the configuration register during the irradiation testing it was found that calculated baseline for the BC2 sometimes changed permanently to another value. As no signal is present on the analogue inputs during irradiation, the calculated baseline value should remain constant as could be seen in the calculated baseline from the BC3. The BC2 is a moving average filter and calculates a new baseline as $baseline = \frac{sum - old\ val + new\ val}{filter\ length}$ where neither the sum or the new and old values are TMR protected. A bit-flip in the sum register would cause the behaviour experienced in the irradiation testing. The simple fix would be to TMR protect the sum register. Another option would be to employ a direct sum calculation instead of a cumulative sum, but as this would add additional adders and multiplexers the space saving compared to using TMR might not be as huge. However, none of these fixes are included in the v3/v4 versions, as these irradiation results were only available after design submission. A workaround for the issue is to employ the auto-reset feature implemented for the SAMPA, which will reset the baseline if the signal stays outside the threshold for longer than a predetermined time.

4.3.5 System Integration Tests

To integrate the SAMPA with the rest of the front-end and readout electronics, the TPC, MCH, STAR, and MPD have all designed prototypes of their final front-end cards, as shown in Fig. 4.12. They have also performed tests of their system to verify that the new front-end cards with SAMPA devices operate as they require them too for the final installation. This includes verifying signal and noise integrity, the proper operation of the slow control, the data readout, triggering and any other parts of the functionality that they require in their system.

The TPC front-end card contains five SAMPA devices running in direct readout mode and connected to two GBTx devices which are further connected to a CRU through optical links. For the TPC, their tests have confirmed that the device operates reliably in combination with the GBTx and GBT-SCA. This includes confirming that the I^2C communication with the GBT-SCA works at full speed with several SAMPAs on the same bus, the multi-drop clock and trigger distribution are operating satisfactorily as well as the GBTx's ability to sample the transmitted signals reliably. Several front-end cards have also been tested when connected to a small TPC chamber in a beam test and has been shown to work reliably in this setting as well. Some optimization has been applied in v3 for the direct readout mode to simplify and make the synchronization phase on start-up more reliable. The primary changes are related to the SAMPAs on/off control of the SLVS drivers when the direct readout mode is enabled, in addition to optimizations in the clock-to-output delay and skew.

The MCH front-end cards are relative small boards, as seen in Fig. 4.12, containing two SAMPA devices. Up to five boards are connected to a long flex-cable connected to a remote concentrator card containing the GBTx and GBT-SCA devices. The testing of the MCH front-end cards proved that the inter-device communication in the daisy chaining functionality works, as well as the communication with the GBTx. It also confirms that it is possible to mount and route the device on boards that are no bigger than their previous boards without greatly affecting noise or signal integrity. Resulting from the MCH testing some optimizations for v3 has been done to avoid packet loss in the daisy chaining link at high occupancies, as well as an issue where a soft reset through the I^2C link would not return an acknowledge due to the device being in process of resetting.

During testing of the I^2C communication between SAMPA and GBT-SCA devices over the long cables for the MCH on the v3 device, issues were discovered with missing acknowledgement of sent information. Due to the cable having a high capacitance of 400 pF and termination not being present, the received signals were misinterpreted by the SAMPA. The termination issues, which sometimes caused a too early response or missing acknowledge due to reflections, was mitigated by adding ferrites to the line and configuring the GBT-SCA to directly drive the SCL line instead of using external pull-up resistors. Subsequently, parts of the issue were still shown to be present where the design was found to sometimes completely stop responding and recovery was only possible by resetting the device. Due to slow rising times caused by the high capacitance of the cable and due to the SAMPA not having Schmitt-

(a) *MCH*

(b) *TPC*

(c) *RHIC STAR*

(d) *NICA MPD*

Fig. 4.12 Detector front-end cards. The SAMPA devices are 15 mm × 15 mm for reference. The MCH board shown is the DS12, their other board, the DS345, has a slightly wider form factor

triggers on the input lines, the input on the SAMPA would incorrectly toggle during a transition, creating incorrectly received data, particularly at the I^2C stop bit. The SAMPA has a built-in watchdog that resets the I^2C module in case it does not receive a new transition on the line within 13 μs, but due to a design error this did not apply to the circuitry that detects the start and stop-messages and the design was stuck waiting for a proper stop message. As the issue was discovered after the final v3 production, it was solved by adding an I^2C buffer (Texas Instruments TCA9803 [31]) in front of the SAMPA to reshape the signal and avoid the issue. The TCA9803 was irradiated with a 200 MeV proton beam to 50 kRad without errors [32].

The STAR front-end card has no special interfacing requirements, but in contrast to the other detectors, it uses a triggered readout mode with a shorter time window. During testing, an issue was found that was only present in the triggered mode, where the memory pointers in the ring buffer came out of sync between the header and data memory, causing incorrect payload data to be sent for a given header. The issue is corrected in v3 and a workaround can be applied for v2 devices.

References

1. Bitvis. (2017) Bitvis utility library. [Online]. Available: http://bitvis.no/products/bitvis-utility-library/.
2. Cadence. (2017) Incisive Enterprise Simulator. [Online]. Available: https://www.cadence.com/content/cadence-www/global/en_US/home/tools/system-design-and-verification/simulation-and-testbench-verification/incisive-enterprise-simulator.html.
3. Mentor. (2017) Questa Advanced Simulator. [Online]. Available: https://www.mentor.com/products/fv/questa/.
4. A. Velure and B. Sanches, *SAMPA Specification*, 2017.
5. S. M. Mahmood, "Exploring Single Event Effects in SAMPA ASIC for Time Projection Chamber (TPC) and Muon CHamber (MCH) detectors in A Large Ion Collider Experiment (ALICE)," Ph.D. dissertation, University of Oslo, Oslo, Norway, 2019, to be published.
6. S. Sutherland and DonMills, *Verilog and SystemVerilog Gotchas*, 2007.
7. W. Snyder. (2017) Verilator. [Online]. Available: https://www.veripool.org/wiki/verilator.
8. Mentor. (2017) HDL Designer. [Online]. Available: https://www.mentor.com/products/fpga/hdl_design/hdl_designer_series/.
9. Cadence. (2017) Incisive Formal Verifier. [Online]. Available: https://www.cadence.com/content/cadence-www/global/en_US/home/tools/system-design-and-verification/formal-and-static-verification/incisive-formal-verification-platform/incisive-formal-verifier.html.
10. Mentor. (2017) FormalPro. [Online]. Available: https://www.mentor.com/products/fv/formalpro/.
11. Cadence. (2017) Conformal Equivalence Checker. [Online]. Available: https://www.cadence.com/content/cadence-www/global/en_US/home/tools/digital-design-and-signoff/equivalence-checking/conformal-equivalence-checker.html.
12. The ALICE Collaboration, "Upgrade of the ALICE read-out and trigger system," CERN, Tech. Design Rep. CERN-LHCC-2013-019, ALICE-TDR-015, 2013. [Online]. Available: http://cds.cern.ch/record/1603472.
13. Cadence. (2018) Modus. [Online]. Available: https://www.cadence.com/content/cadence-www/global/en_US/home/tools/digital-design-and-signoff/test/modus-test.html.
14. C. E. Cummings, "Clock domain crossing (CDC) design & verification techniques using SystemVerilog," in *Synopsis User Group Meeting*, 2008.

15. Mentor. (2017) Questa CDC. [Online]. Available: https://www.mentor.com/products/fv/questa-cdc/.
16. Terasic. (2017) Altera SoCKit - the Development Kit for New SoC Device. [Online]. Available: http://www.terasic.com.tw/cgi-bin/page/archive.pl?Language=English&CategoryNo=205&No=816&PartNo=1.
17. Altera, *Cyclone V Device Overview*, June 2016, Version: 2016.06.10. [Online]. Available: https://www.altera.com/en_US/pdfs/literature/hb/cyclone-v/cv_51001.pdf.
18. JEDEC, *JESD8-13 - Scalable Low-Voltage Signaling for 400 mV (SLVS-400)*, Std., oct 2001.
19. B. Huang, R. Vann, S. Freethy, R. Myers, G. Naylor, R. Sharples, and V. Shevchenko, "FPGA-based embedded Linux technology in fusion: The MAST microwave imaging system," *Fusion Engineering and Design*, vol. 87, no. 12, pp. 2106–2111, dec 2012.
20. S. Korolczuk, S. Mianowski, J. Rzadkiewicz, P. Sibczynski, L. Swiderski, and I. Zychor, "Digital acquisition in high count rate gamma-ray spectrometry," *IEEE Transactions on Nuclear Science*, vol. 63, no. 3, pp. 1668–1673, June 2016.
21. F. Costa, V. C. Barroso, C. Soos, R. Divia, A. Wegrzynek, T. Kiss, H. Engel, G. Simonetti, J. Niedziela, B. V. Haller, A. Telesca, F. Carena, P. V. Vyvre, U. Fuchs, S. Chapeland, and W. Carena, "The ALICE C-RORC GBT card, a prototype readout solution for the ALICE upgrade," in *2016 IEEE-NPSS Real Time Conference (RT)*, jun 2016, pp. 1–5.
22. M. Pesaresi, M. B. Marin, G. Hall, M. Hansen, G. Iles, A. Rose, F. Vasey, and P. Vichoudis, "The FC7 AMC for generic DAQ & control applications in CMS," *Journal of Instrumentation*, vol. 10, no. 03, p. C03036, mar 2015.
23. R. Brun and F. Rademakers, "ROOT — an object oriented data analysis framework," *Nuclear Instruments and Methods in Physics Research Section A: Accelerators, Spectrometers, Detectors and Associated Equipment*, vol. 389, no. 1, pp. 81–86, apr 1997.
24. Intel, *Avalon Interface Specifications*, Std., 2018.
25. J. Dugan, S. Elliott, B. A. Mah, J. Poskanzer, and K. Prabhu. (2016) iPerf - The network bandwidth measurement tool. [Online]. Available: https://iperf.fr/.
26. D. de Carvalho, B. Sanches, M. D. Carvalho, and W. V. Noije, "A flexible stand-alone FPGA-based ATE for ASIC manufacturing tests," in *2018 IEEE 19th Latin-American Test Symposium (LATS)*, March 2018, pp. 1–6.
27. S. Barboza, M. Bregant, V. Chambert, B. Espagnon, H. H. Herrera, S. Mahmood, D. Moraes, M. Munhoz, G. Noël, A. Pilyar, P. Russo, B. Sanches, G. Tambave, K. Tun-Lanoë, W. van Noije, A. Velure, S. Vereschagin, T. Weber, and S. Zaporozhets, "SAMPA chip: a new ASIC for the ALICE TPC and MCH upgrades," *Journal of Instrumentation*, vol. 11, no. 02, p. C02088, feb 2016.
28. G. Tambave and A. Velure, "Qualification of the ALICE SAMPA ASIC with a high-speed continuous DAQ system," *IEEE Transactions on Nuclear Science*, vol. 64, no. 6, pp. 1461–1466, jun 2017.
29. J. Adolfsson, A. A. Pabon, M. Bregant, C. Britton, G. Brulin, D. Carvalho, V. Chambert, D. Chinellato, B. Espagnon, H. H. Herrera, T. Ljubicic, S. Mahmood, U. Mjörnmark, D. Moraes, M. Munhoz, G. Noël, A. Oskarsson, L. Osterman, A. Pilyar, K. Read, A. Ruette, P. Russo, B. Sanches, L. Severo, D. Silvermyr, C. Suire, G. Tambave, K. Tun-Lanoë, W. van Noije, A. Velure, S. Vereschagin, E. Wanlin, T. Weber, and S. Zaporozhets, "SAMPA chip: the new 32 channels ASIC for the ALICE TPC and MCH upgrades," *Journal of Instrumentation*, vol. 12, no. 04, pp. C04 008–C04 008, apr 2017.
30. S. M. Mahmood, K. Røed, A. Velure, and F. L. Winje, "First Irradiation Test Results of the ALICE SAMPA ASIC," in *Topical Workshop on Electronics for Particle Physics*, 2017.
31. Texas Instruments, *TCA9803 Datasheet*, 2017. [Online]. Available: www.ti.com/product/TCA9803.
32. A. Ferrero, "Readout chain validation," MCH Solar PRR Meeting, 2018. [Online]. Available: https://indico.cern.ch/event/728037/contributions/2998040/attachments/1654237/2649964/SOLAR-PRR-20180525-readout_chain_tests.pdf.

Chapter 5
Conclusion and Outlook

5.1 Conclusion

A major upgrade of the ALICE experiment will be realised in 2019/2020. The goal is to be able to inspect all heavy-ion collisions at the full expected LHC luminosity of $6 \times 10^{27}\,\text{cm}^{-2}\,\text{s}^{-1}$ at 50 kHz interaction rate. The SAMPA ASIC has been designed to support a new, faster readout of the Time Projection Chamber (TPC) and Muon Tracking Chamber (MCH) detectors. The third version of the device has passed the production readiness review for both TPC [1] and MCH [2] and so fulfils the requirements set up by the detectors and the design can be considered as completely functional according to specifications.

The device has also been evaluated in test beams with partial detector setups. Figure 5.1 shows the result of a noise measurement, by the MCH, of one quadrant with 640 channels (10 FECs, 20 SAMPA devices) and the recorded beam profile of the 120 GeV Super Proton Synchrotron (SPS) beam used for testing [3]. Figure 5.2 shows the TPC noise and pedestal test results for 960 channels (6 FECs, 30 SAMPA devices) and Fig. 5.3 shows a particle crossing the same chamber [4].

The amount of time spent on verification versus design is usually taken to be between 40 and 70%. If we base our analysis on the amount of code written, we can see in Table 5.1 that we are at the upper range of this level for the SAMPA design. A few issues were found during functional testing of the v2 device and the testing effort was stepped up for verification of the v3 design. As only one hard to detect issue[1] has been discovered for the v3 device the design could be considered to be verified sufficiently. The design and verification effort has passed an engineering design review [5], as well as a production readiness review [6].

Due to the changes in specifications from the early concepts to the final design, the device is more complex than what is currently needed for the two primary detectors.

[1] See Sect. 4.3.5. That part of the design was borrowed from CERN and so is also likely present in several other CERN devices.

A. Velure, *A Digital Signal Processor for Particle Detectors*, Springer Theses,
https://doi.org/10.1007/978-3-030-71559-5_5

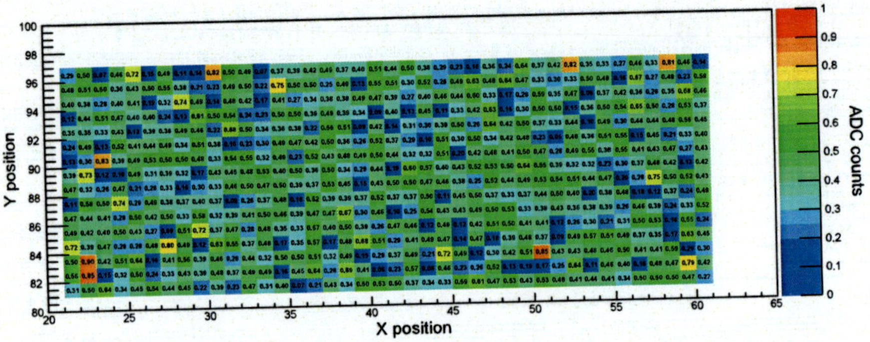

(a) *Quadrant 1 noise measurement*

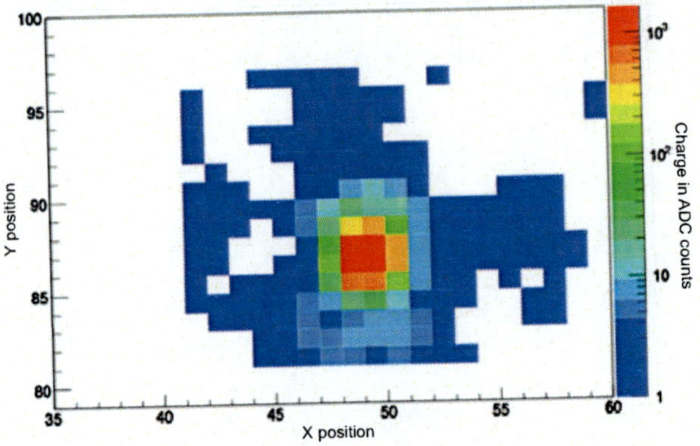

(b) *Quadrant 1 beam profile measurement*

Fig. 5.1 Noise and beam profile measurement with quadrant 1 of the MCH detector [3]

Table 5.1 Amount of code as counted by Cloc [7]

Source	Files	Comments	Code lines
SAMPA	53	1337	8749
Testbenches	35	2841	16000

The complexity of the design also caused unforeseen delays in the development process, which put strains on the time allotted for final production, testing and commissioning. If this had been originally anticipated, it might have been seen that it would have been both cheaper and better to produce a device that was simpler in design, only containing the existing analogue front-end, the ADCs, and a small amount of digital logic that takes care of configuring the device and directly serializing the ADC data. This device would satisfy the TPC requirements and operate in practically the same way as the TPC is currently using the SAMPA. A secondary device, be it an FPGA or another ASIC, could do the DSP functionality for the detectors that require this functionality.

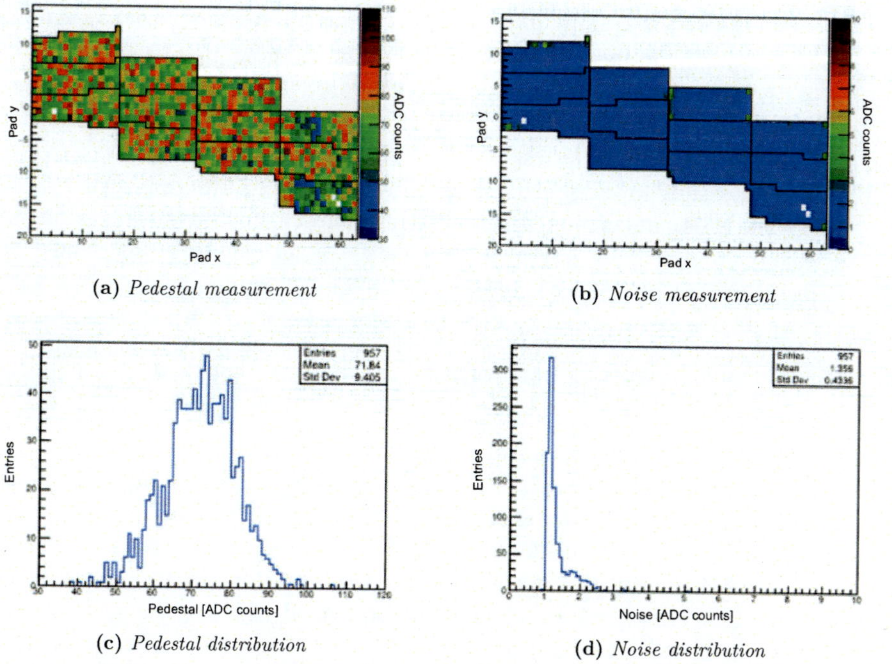

(a) *Pedestal measurement*

(b) *Noise measurement*

(c) *Pedestal distribution*

(d) *Noise distribution*

Fig. 5.2 Pedestal and noise measurement of one TPC inner readout chamber [4]

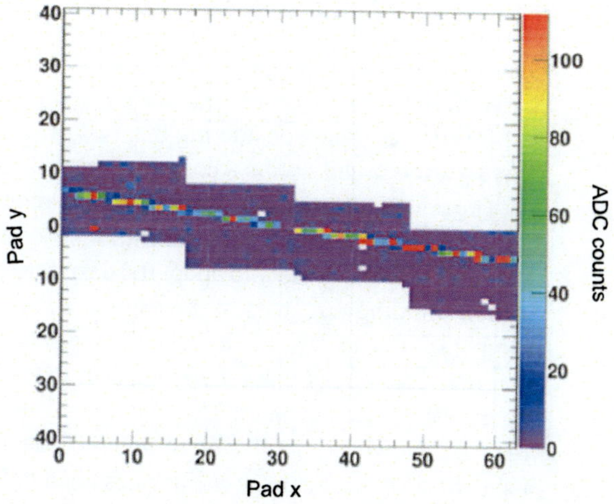

Fig. 5.3 Particle track crossing the TPC chamber [4]

Going with a simplified front-end design would not likely have saved more than a few months in the SAMPA design process, as the digital design was not one of the bottlenecks. However, it would have reduced the time needed to test each device by several minutes in the final production test, which saves some additional months since there is a need to test 80 000 devices. A smaller device also provides better yield, possibly lowering the number of wafers, and so lowering costs.

The DSP part benefits from processing many channels as resources can be shared and so the size of the design can be reduced. With a simplified front-end device, this benefit is less pronounced, which presents the option of going back to a design with only 16 channels. This simplifies the analogue layout work and reduces channel-to-channel variations. This could also have significantly reduced the analogue design time and provided a device at an earlier date, as well as giving even better yields at the cost of occupying more board area. However, a 16-channel device is a disadvantage for the MCH as their board area is sparse.

In the case of the MCH, where the area occupied by the devices and its power consumption is a counter argument to using FPGAs, it would have been possible to design a separate custom dedicated ASIC only for the digital processing functionality. As the MCH currently only uses a small part of the functionality of the SAMPA, the size of the simplified front-end device combined with a new device that replicates the needed MCH functionality would still likely be smaller and consume less power than the current device. The primary area and power consumption of the current device origins from the large buffer memories originally required by the TPC. Since the MCH only needs 64 channels per board, the device could have been built to accept data from two 32-channel devices or four 16-channel devices. This would have saved the effort of developing and testing the daisy chaining implementation as well. In total, the development time could likely have been reduced to two years for this device. But, considering that two different devices need to be manufactured, this solution might not be a cheaper option than the SAMPA.

The other detectors, apart from the TPC and MCH, that are interested in the device and require more compression or filtering capabilities could utilize an FPGA, which provides more flexibility to adapt the design to their present needs and gives them an option of doing upgrades of the firmware code in the future. Additionally, it opens the possibility to offload the data through the use of high speeds transceivers that can reach data rates that are higher then what is available with the GBTx. This has an additional advantage of lowering the needed number of optical fibres, as well as the cost. However, the use of commercial devices in an irradiated area will always have the disadvantage that the devices need to be qualified for operation and that the firmware code for the FPGA needs to be designed with SEE effects in mind.

It was decided not to use Huffman compression with differential encoding for the TPC readout, as the compression factor was believed to be too much dependent on detector properties such as noise, baseline instabilities and gain variations compared to conventional zero suppression [8]. This is in essence true, in the general sense [9], but it strongly depends on the signal properties of the detector, which were not known at the time of analysis. The strength of the zero suppression is based on a stable baseline, which requires the baseline correction functionality to be enabled.

On the other hand, nothing prevents one from employing the baseline correction also for the Huffman to increase the compression. However, the output data will no longer be lossless, but it will provide more data for analysis than the zero suppression as data between clusters are not lost. Another objection to the use of Huffman was the required resources needed in the receiving CRU FPGA to do the decoding of all the front-end cards originally envisioned to be connected to one CRU [10]. The tested implementation is, however, not optimal and can be improved upon, though it will still use more resources than what is needed for processing zero suppressed data or raw data as in the final implementation. A lack of resources can be alleviated by adding more CRUs, but this comes down to cost versus quality of results. The final system for the TPC reduced the sampling rate from 10 to 5 MHz and increased the number of links from four to eleven, and decreased the link speed to 160 Mbps. With some analysis of the new detector data to confirm the stability of the signal baseline, it would be possible to switch the sampling back to using 10 MHZ, and by employing Huffman, be able to transfer the higher sampled data over the same links.

As it stands, the strength of the SAMPA lies in its versatility, high flexibility, and configurability, particularly of the digital design, which makes it applicable to a wide range of gas-based detectors. Its limiting factors are primarily the restricted selection of analogue gain and shaping times available.

5.2 Outlook

Preparation of the mass-production testing for the devices that eventually will be used for the upgrade is in the works. The test plan for detection of manufacturing defects is expected to be the same as used for the previous tests as described in Sect. 4.3.2.1. The test-framework will though need to be ported to another platform that can support more devices per DAQ unit.

The ALICE experiment uses a framework called AliRoot [11] for their simulation, reconstruction, and analysis. The framework is able to simulate the full detector performance including the readout performance and so in the future, there would be a need to implement a sufficiently faithful model of the SAMPA as well. An option that has been investigated, but not evaluated, is to use a program called Verilator [12] that automatically ports Verilog code to SystemC code. This would simplify the conversion process by avoiding the need to have a deep knowledge of the design, but might increase in terms of simulation speed.

The DAQ system was designed with the limitations of the chosen FPGA evaluation board, the need for testing all digital functionality of the design, and the limited time frame for the development work. If a detector chooses to base its acquisition system around the design, some considerations should be taken. For operating with the SAMPA data links at 320 Mbps, a faster device should be chosen, as the design currently uses extensive optimization techniques to reach timing, which complicates the design and makes it hard to maintain. For operating at slower speeds, the design can be simplified, as was done in the MCH testing. To increase data throughput, it would be beneficial to implement the Ethernet communication directly in firmware, instead of relying on the Linux system to handle this.

The evaluation board used for the DAQ system, the Altera SocKit [13], is out of production, and is replaced by a similar board, the Altera DE10 [14]. The new board does not support high-speed transceivers on the connector connecting to the SAMPA test board. As such, the SFP transceivers on the test board cannot be used with this evaluation board. An effort to port the SocKit design to the DE10 is nearly complete, but requires some final debugging.

The current fixed shaping and gain settings were specifically chosen for the needs of the TPC and MCH, but they might not be suitable for other detectors that wish to utilize the SAMPA. Therefore, an alteration to the design, implementing programmable gain and shaping settings, could be done to make the device more general. As this adds more circuitry to the sensitive analogue section, it comes at the cost of increased noise.

To improve manufacturing tests and calibration, a method for automatically injecting a known charge into the front-end channels would be of interest. A switching network could be used to connect one or more capacitors with known charge onto each channel. A digital to analogue converter was designed for this purpose, but it was not implemented due to the risks that it would increasing the noise.

Throughout Chap. 3 and in Sect. 4.1.4 there are discussions on some additional solutions and improvements in the digital design that might prove useful to include if a new device is designed, depending on the needs of the detectors involved. As the digital design was produced with enough flexibility so that it could be used for a wide variety of detector applications, it compromises on device complexity and power consumption. If this is of concern for a detector application, the design is built with enough modularity so that removing unneeded blocks in a new design can be done without too much effort.

References

1. A. Kluge *et al.* (2018) TPC FEC PRR. [Online]. Available: https://indico.cern.ch/event/721919/.
2. A. Kluge *et al.* (2018) MCH Dual Sampa PRR. [Online]. Available: https://indico.cern.ch/event/728720/.
3. A. Baldisseri, "MCH 2018 SPS testbeam," MCH Upgrade Weekly Meeting, 2018. [Online]. Available: https://indico.cern.ch/event/762022/contributions/3162180/attachments/1726348/2788778/MCH-Sept-SPS2018Testbeam-Oct218.pdf.
4. H. Schulte, "Test beam may 2018: Pedestals, noise and pulser analysis," TPC upgrade collaboration meeting, 2018. [Online]. Available: https://indico.cern.ch/event/653116/contributions/2701891/attachments/1513895/2362265/Testbeam_may_2017_-_pedestals_and_noise.pdf.
5. A. Kluge *et al.* (2017) Sampa engineering design review. [Online]. Available: https://indico.cern.ch/event/617831/.
6. A. Kluge *et al.* (2018) Sampa PRR. [Online]. Available: https://indico.cern.ch/event/700894/.
7. A. Danial. (2017) Count lines of code. [Online]. Available: https://github.com/stsnel/cloc.

8. H. Appelshäuser, M. Arslandok, M. Ivanov, C. Lippmann, and J. Wiechula, “Readout scheme of the upgraded ALICE TPC,” CERN, Tech. Rep. ALICE-PUBLIC-2016-006, Nov 2016. [Online]. Available: https://cds.cern.ch/record/2231785.
9. S. Klewin, “Length-limited Huffman coding,” TPC CRU Meeting, 2015. [Online]. Available: https://indico.cern.ch/event/453573/contributions/1951400/attachments/1169252/1687190/klewin_2015_10_13.pdf.
10. S. Klewin, “Huffman decoder,” TPC CRU Meeting, 2015. [Online]. Available: https://indico.cern.ch/event/462008/contributions/1978677/attachments/1205143/1755806/klewin_2015_12_15.pdf.
11. The ALICE Collaboration. (2017) AliRoot - the ALICE offline software. [Online]. Available: http://alice-offline.web.cern.ch/.
12. W. Snyder. (2017) Verilator. [Online]. Available: https://www.veripool.org/wiki/verilator.
13. Terasic. (2017) Altera SoCKit - the Development Kit for New SoC Device. [Online]. Available: http://www.terasic.com.tw/cgi-bin/page/archive.pl?Language=English&CategoryNo=205&No=816&PartNo=1.
14. Terasic. (2019) Altera DE10-Standard. [Online]. Available: http://www.terasic.com.tw/cgi-bin/page/archive.pl?No=1081.

Appendix A
Schematic Legends

Some simplifications have been done in the schematic drawings to avoid redundancy and make them easier to read. The custom/uncommon symbols are shown in Fig. A.1.

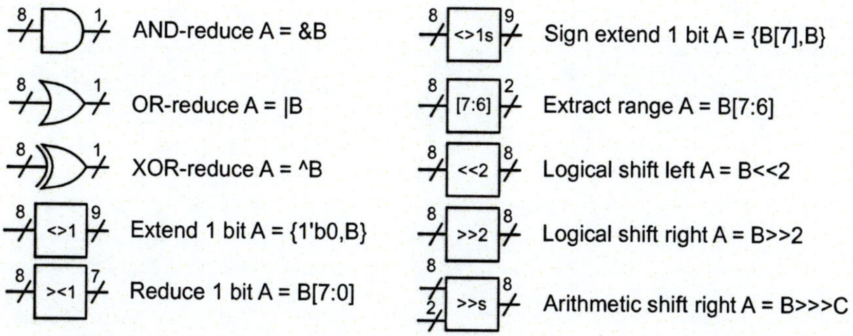

Fig. A.1 Schematic legend

A. Velure, *A Digital Signal Processor for Particle Detectors*, Springer Theses, https://doi.org/10.1007/978-3-030-71559-5

Appendix B
Registers

The SAMPA contains two types of registers; global and channel. The global registers are directly accessible through the I^2C interface and contain configuration that concerns the main operation of the device or configurations that are common to all channels. The channel registers exist in each channel and contain configuration that can be individually set for each channel. The channel registers are only accessible via the global register, as explained in Sect. B.1.3.

B.1 Global Registers

The full list of global registers is listed in Tables B.1 and B.2. Following is the extended description of the registers.

B.1.1 Pin Status

These registers show the status of input pins on the SAMPA.

HWADD (Hardware address) The device address. It is used both as the device's I^2C address as well as a source indicator in the header of the data packets sent to the readout unit.

CLKCONF (Clock configuration) Status of the clk_config pins used for controlling the internal clock division.

BOUNDARY Gives the status of the trigger inputs and the daisy chaining pins.

A. Velure, *A Digital Signal Processor for Particle Detectors*, Springer Theses,
https://doi.org/10.1007/978-3-030-71559-5

Table B.1 Global registers part 1

Register name		Address	Type	Default	Description	
HWADD	[3:0]	0x00	R	0x00	[3:0]	Chip address (hardware address)
TRCNTL	[7:0]	0x01	R	0x00	[7:0]	Trigger count, lower byte
TRCNTH	[7:0]	0x02	R	0x00	[7:0]	Trigger count, upper byte
BXCNTLL	[7:0]	0x03	R	0x00	[7:0]	Bunch crossing count, lower byte
BXCNTLH	[7:0]	0x04	R	0x00	[7:0]	Bunch crossing count, mid byte
BXCNTHL	[3:0]	0x05	R	0x00	[3:0]	Bunch crossing count, upper byte
PRETRG	[7:0]	0x06	RW	0x00	[7:0]	Number of pre-samples (Pre-trigger delay), max 192
TWLENL	[7:0]	0x07	RW	0xE7	[7:0]	Number of cycles for time window +1, lower byte
TWLENH	[1:0]	0x08	RW	0x03	[1:0]	Number of cycles for time window +1, upper byte
ACQSTARTL	[7:0]	0x09	RW	0x00	[7:0]	Number of cycles to wait before acquisition starts, lower byte
ACQSTARTH	[1:0]	0x0A	RW	0x00	[1:0]	Number of cycles to wait before acquisition starts, upper byte
ACQENDL	[7:0]	0x0B	RW	0xFF	[7:0]	Number of cycles elapsed from trigger to acquisition end +1, lower byte
ACQENDH	[1:0]	0x0C	RW	0x03	[1:0]	Number of cycles elapsed from trigger to acquisition end +1, upper byte
VACFG	[7:0]	0x0D	RW	0x30		Various configuration settings
			RW	0x00	[0]	Continuous mode enabled
			RW	0x00	[1]	Raw data enable
			RW	0x00	[2]	Cluster sum enable
			RW	0x00	[3]	Huffman enable
			RW	0x01	[4]	Enable header generation for empty channels
			RW	0x01	[5]	Power save enable
			RW	0x00	[6]	Enable automatic clock gating on I^2C block
			RW	0x00	[7]	Enable clock gating on neighbour block when number of neighbour is 0
CMD	[2:0]	0x0E	RW	0x00	[2:0]	Commands, see Table B.3

(continued)

Table B.1 (continued)

Register name		Address	Type	Default	Description	
NBCFG	[7:0]	0x0F	RW	0x40		Neighbour configuration settings
			RW	0x00	[5:0]	Neighbour input delay, ca. 0.2 ns per bit for a total of ca. 12.5 ns
			RW	0x01	[7:6]	Number of neighbours
ADCDEL	[6:0]	0x10	RW	0x00		ADC sampling clock delay
			RW	0x00	[5:0]	ADC sampling clock delay, ca. 1.5 ns per bit for a total of ca. 94.5 ns
			RW	0x00	[6]	Invert ADC sampling clock
ADCTRIM	[2:0]	0x11	RW	0x04	[2:0]	Voltage reference trimming
SOCFG	[5:0]	0x12	RW	0x34		Serial link configuration
			RW	0x04	[3:0]	Number of serial out, 0-11
			RW	0x01	[4]	Disable internal termination of input differential links
			RW	0x01	[5]	Enable NBflowstop_in pin
SODRVST	[7:0]	0x13	RW	0x55		Serial link drive strength configuration, see Table B.5
			RW	0x01	[1:0]	Drive strength of serial out 4-0
			RW	0x01	[3:2]	Drive strength of neighbour flow stop out/serial out 5
			RW	0x01	[5:4]	Drive strength of serial out 6,8,10
			RW	0x01	[7:6]	Drive strength of serial out 7,9

B.1.2 Event Management

These registers show the configuration options for setting up events.

TRCNTL, TRCNTH (Trigger count) The number of triggers that have been processed by the device. The count is increased for each new trigger, regardless of if the trigger source is the event trigger pin, the soft trigger command or from running in continuous mode. The counter can be reset with the TRCLR soft command, see Table B.3. The counter will automatically roll around after 2^{16} triggers if it is not manually reset. The value is not used for any internal operational purpose.

Table B.2 Global registers part 2

Register name		Address	Type	Default	Description	
ERRORS	[7:0]	0x14	R	0x00		Errors accumulated
			R	0x00	[4:0]	Correctable header hamming errors
			R	0x00	[7:5]	Uncorrectable header hamming errors
PMADDL	[7:0]	0x15	RW	0x00	[7:0]	Pedestal memory address, lower byte
PMADDH	[1:0]	0x16	RW	0x00	[1:0]	Pedestal memory address, upper byte
CHRGADD	[4:0]	0x17	RW	0x00	[4:0]	Channel register address
CHRGWDATL	[7:0]	0x18	RW	0x00	[7:0]	Channel register write data, lower byte
CHRGWDATH	[4:0]	0x19	RW	0x00	[4:0]	Channel register write data, upper byte
CHRGCTL	[7:0]	0x1A	RW	0x00		Channel register control
			RW	0x00	[4:0]	Channel number
			RW	0x00	[5]	Broadcast to all channels (channel number ignored)
			RW	0x00	[6]	Write, not read from register address (returns to read after write)
			RW	0x00	[7]	Increment PMADD (returns automatically to zero)
CHRGRDATL	[7:0]	0x1B	R	0x00	[7:0]	Channel register read data, lower byte
CHRGRDATH	[4:0]	0x1C	R	0x00	[4:0]	Channel register read data, upper byte
CHORDAT	[4:0]	0x1D	RW	0x00	[4:0]	Channel readout order data
CHORDCTL	[5:0]	0x1E	RW	0x00		Channel readout order control
			RW	0x00	[4:0]	Position in order
			RW	0x00	[5]	Write enable
BYPASS	[3:0]	0x1F	RW	0x00	[3:0]	Bypass inputs to serial 0, see Table B.6
SERCHSEL	[4:0]	0x20	RW	0x00	[4:0]	Channel select for ADC test serializer mode in bypass
RINGCNT	[7:0]	0x21	R	0x00	[7:0]	Ring oscillator counter difference from reference ADC clock

(continued)

Table B.2 (continued)

Register name		Address	Type	Default	Description	
CLKCONF	[6:0]	0x22	R	0x00	[6:0]	Clock configuration pin status
BOUNDARY	[4:0]	0x23	R	0x00		Status of differential input pins
			R	0x00	[0]	NBflowstop_in
			R	0x00	[1]	DinN
			R	0x00	[2]	hb_trg
			R	0x00	[3]	trg
			R	0x00	[4]	bx_sync_trg
CHEN0	[7:0]	0x24	RW	0xFF	[7:0]	Channel enable 7-0
CHEN1	[7:0]	0x25	RW	0xFF	[7:0]	Channel enable 15-8
CHEN2	[7:0]	0x26	RW	0xFF	[7:0]	Channel enable 23-16
CHEN3	[7:0]	0x27	RW	0xFF	[7:0]	Channel enable 31-24

Table B.3 Command register, register returns to 0 after command is executed

Command name	CMD[2:0]	Description
NOP	0x0	No operation
SWTRG	0x1	Software trigger
TRCLR	0x2	Clear trigger counter
ERCLR	0x3	Clear errors
BXCLR	0x4	Clear bunch crossing counter
SOFTRST	0x5	Software reset
LNKSYNC	0x6	Generates sync packet on serial links
RINGOSCTST	0x7	Run ring oscillator test

BXCNTLL, BXCNTLH, BXCNTHL (Bunch crossing count) The current value of the bunch crossing counter. The counter runs on the LHC 40 MHz bunch-crossing clock (internally generated or externally provided) and its value is captured at the start of each processing time window and added to the header of the corresponding packet for that processing time window. It is also added to the heartbeat packet when one is requested through pulsing of the heartbeat trigger pin (hb_trg). The counter can be reset with the BXCLR soft command, see Table B.3, or by pulsing the bx_sync_trg pin. If not manually reset, it will automatically roll around after 2^{20} cycles.

PRETRG (Pre-trigger) Sets the delay of the input data in relation to the trigger signal. A one sample-delay is equivalent to the trigger signal arriving one ADC clock cycle earlier. The configuration is intended to compensate for delays in the external trigger signal or to give more time for trigger decision. The maximum configurable delay is 192 samples which correspond to 19.2 μs when running at an ADC clock of 10 MHz. The delay is inserted at the entry of the digital design before any data handling or compression happens. The configuration does not alter the length of the time window, but merely moves the trigger point in the sample stream earlier in time than when the trigger was received.

TWLENL, TWLENH (Time window length) The number of cycles +1 that a time window will last from a trigger start. In continuous mode, this will impact the average length of packets.

ACQSTARTL, ACQSTARTH (Acquisition start) The number of cycles from the start of a trigger that will be suppressed in the output data when running in zero-suppressed mode or cluster sum mode. It is useful for instance in cases where it is known that there always will be an unwanted signal perturbation directly after the trigger which could badly influence the filters. Set the value to 0 to disable.

ACQENDL, ACQEND (Acquisition end) The number of cycles from the start of trigger to when sample acquisition should end. The remaining samples after this point, until the end of the time window, will be suppressed when operating in zero suppression mode or cluster sum mode. If the value is set higher or equal to TWLEN then it has no effect. The boolean equation is $acquisition\_enable = (TWLEN >= ACQSTART)\&(TWLEN <= ACQEND)$.

VACFG[0] Continuous mode enable Enables automatic generation of a new trigger at the end of a time window. A manual soft or pin trigger would prematurely end the current time window and start a new one. Manual pin triggering can be used to align time windows across multiple SAMPAs since they will all start at the same time.

CMD:TRCLR (Trigger clear) Clears the trigger counter TRCNT.

CMD:SWTRG (Software trigger) Provides a single soft trigger, mainly for test purposes.

CMD:BXCLR (Bunch crossing clear) Clears the bunch crossing counter BXCNT.

CHENx (Channel enable) Provides the possibility to shut down specific channels to avoid that they send data. It works by disabling all the filters and suppressing the data readout for the channel in addition to disabling the clock for the ADC.

B.1.3 Channel Register Access

The register layout has been designed to minimize the number of writes needed to update the channel registers and pedestal memories of each channel. By using the I^{2}C automatic address increment feature, it is possible to complete all writes needed to update one channel or pedestal address in one continuous I^{2}C command. An additional broadcast bit can be set to write the same data to all channels so that individual access is not needed. When filling the pedestal memory it is possible to set the pedestal memory to increment on each write avoiding the need to update the two registers each time.

Write

1. Set CHRGADD (Channel register address) to the address of the channel register that you wish to write to.
2. Set the data to write at CHRGWDATH:CHRGWDATL (Channel register write data).
3. Set CHRGCTL[6] (Channel register control) high for write, set CHRGCTL[5] high if you wish to write the same value to all channels (broadcasting), set CHRGCTL[4:0] to the channel number that you wish to write the data to. When broadcasting the channel number is ignored.

Read

1. Set CHRGADD to the address of the channel register that you wish to read from.
2. Set CHRGCTL[6] low for read, set CHRGCTL[4:0] to the channel number that you wish to read from. Broadcast (CHRGCTL[5]) is ignored for reads.
3. The data will appear at CHRGRDATH:CHRGRDATL (Channel register read data).

Pedestal memory write

1. Make sure the data path configuration for the channel to be written to (DPCFG, see Table B.9) is not using a lookup function f( ), as the lookup function utilizes the pedestal memory and will cause corrupted writes.
2. Set PMADDH:PMADDL (Pedestal memory address) to the address in the pedestal memory that you wish to write to.
3. Set CHRGADD to 0x10 which is the address for PMDATA (Pedestal memory data) in the channel register.
4. Set the data to write at CHRGWDATH:CHRGWDATL.
5. Set CHRGCTL[7] high to automatically increment the currently set pedestal memory address (increment before write), set CHRGCTL[6] high for write, set CHRGCTL[5] high if you wish to write the same value to all channels (broadcasting), set CHRGCTL[4:0] to the channel number that you wish to write to. When broadcasting the channel number is ignored.

Pedestal memory read

1. Make sure the data path configuration for the channel to be read from (DPCFG, see Table B.9) is not using a lookup function f( ), as the lookup function utilizes the pedestal memory and will cause corrupted reads.
2. Set the address in the pedestal memory that you wish to read from at PMADDH:PMADDL.
3. Put the register address for the channel register PMDATA (0x10) at CHRGADD.
4. Set CHRGCTL[7] high to automatically increment the currently set pedestal memory address (increment before read), set CHRGCTL[6] low for read, set CHRGCTL[4:0] to the channel that you wish to read from. Broadcast (CHRGCTL[5]) is ignored for reads.
5. The data will appear at CHRGRDATH:CHRGRDATL.

B.1.4 Channel Ordering

The channel ordering is kept as an ordered list of channels. The order of the channels in the list defines the order that the channels will be read out in. Subsequently, it also defines the serial link that a channel will be read out on, see Table B.4. Note that reading and writing to CHORDAT does not necessarily talk to the same register. When you read from CHORDAT, you read directly from the register that CHORDCTL[4:0] is pointing to. Writing to CHORDAT writes to a temporary register and the data in this register is only moved to the address that CHORDCTL[4:0] is pointing to when the write command CHORDCTL[5] is set high.

Write

1. Set the channel number that you want to place in the list at CHORDAT (Channel ordering data).
2. Set CHORDCTL[5] (Channel ordering control) high for write and set CHORDCTL[4:0] to the position in the list that you want to place the channel in.

Read

1. Set CHORDCTL[4:0] to the position in the list that you want to read from.
2. The data will appear at CHORDAT.

B.1.5 ADC Configuration

These registers enable configuration of some aspects of the ADC operation.

ADCDEL (ADC delay) As shown in Fig. 3.5 there is available a possibility to delay the clock supplied to the ADCs (ckoutADCanalog) in relation to the clock for the digital part (ckoutADC). This feature could be used to reduce the influence of the digital part switching noise on the ADC sampling. The phase can be changed by with a selection of ca. 1.5 ns up to ca. 94.5 ns depending on process variations, which amounts to almost a full cycle at the minimum designed for ADC clock speed of 10 MHz. If bit [6] is enabled the clock is also inverted, which gives a less process dependent phase shift.

ADCTRIM (ADC trimming) Sets the configuration for the Configurable Reference Voltage Source for the analogue front-end which adjusts the $v450$, $v600$ and $v750$.

B.1.6 Serial link configuration

These registers enable configuration of the serial link driver and receivers.

NBCFG (Neighbour configuration)

5:0 Input delay As the serial link speed from the neighbouring slave device in a daisy chain runs at the same speed as the serial link clock on the master device, there is no faster clock available which can be used for oversampling the input. So to avoid metastability on the input a delay chain is provided to encompass this. The delay is process dependent, but in most cases, only an approximate delay is needed which leaves room for the process variations. The delay can be changed with a selection of ca. 0.2 ns up to ca. 12.5 ns, which amounts to a full cycle at the minimum designed for serial link clock speed of 80 MHz.

7:6 Neighbour priority Sets the priority of the neighbour input data in relation to the internal data. It should be set to the number of devices following the current device in the chain. If it is the first device in the chain and there are two following, it should be set to two. If it is the second device in the chain and there is one following, it should be set to one, etc. If the value is set to two, then the device will take two packets from the neighbouring link for each packet it takes from any of the internal channels. Only serial link 0 can output data from the neighbouring link.

SOCFG (Serial out configuration)

3:0 Active links Sets the number of active serial link outputs. The links are activated in sequential order, setting the value to 0 disables all links. Disabling a link also disables its corresponding driver to save power. See Table B.4 for which channel would be connected to which link when this setting is changed.

4 Input termination The physical connections to the input SLVS links are assumed to be differential transmission lines with a differential impedance of 100 Ω. The receivers have a common built-in 100 Ω terminations, which can be enabled if required.

5 NBflowstop_in enable The NBflowstop_in pin should be pulled low if not in use. Setting this bit to zero will set the pin to zero internally so that serial outputs will work normally in case the pin externally is stuck at one.

SODRVST (Serial out drive strength) The serial link drivers have a programmable output current. This enables power saving when transmitting over short distances or it provides the possibility to compensate for the loss in long cables. The configurations are listed in Table B.5.

CMD:LINKSYNC (Link sync) Forces sending of a sync packet on all active links after the current transmission on the link has completed.

B.1.7 Data Compression

Several methods of compressing the data stream before transmission is available. If no method is selected it defaults to zero suppression. In case more than one method is selected, the priority is raw > Huffman > cluster sum > zero suppression.

Table B.4 The table lists on which serial link a channel will be present when the configured number of serial links are set. When the channel re-ordering is in use, then the number in the channel column represent the channel re-ordering list position

Channel	Number of serial links enabled										
	1	2	3	4	5	6	7	8	9	10	11
0	0	0	0	0	0	0	0	0	0	0	0
1	0	0	0	0	0	0	0	0	0	0	0
2	0	0	0	0	0	0	0	0	0	0	0
3	0	0	0	0	0	0	0	0	0	0	1
4	0	0	0	0	0	0	0	1	1	1	1
5	0	0	0	0	0	0	1	1	1	1	1
6	0	0	0	0	0	1	1	1	1	1	2
7	0	0	0	0	1	1	1	1	1	1	2
8	0	0	0	1	1	1	1	2	2	2	2
9	0	0	0	1	1	1	1	2	2	2	3
10	0	0	0	1	1	1	2	2	2	2	3
11	0	0	1	1	1	1	2	2	2	3	3
12	0	0	1	1	1	2	2	3	3	3	4
13	0	0	1	1	1	2	2	3	3	3	4
14	0	0	1	1	2	2	2	3	3	4	4
15	0	0	1	1	2	2	3	3	3	4	5
16	0	1	1	2	2	2	3	4	4	4	5
17	0	1	1	2	2	3	3	4	4	5	5
18	0	1	1	2	2	3	3	4	4	5	6
19	0	1	1	2	2	3	3	4	4	5	6
20	0	1	1	2	3	3	4	5	5	6	6
21	0	1	1	2	3	3	4	5	5	6	7
22	0	1	2	2	3	4	4	5	5	6	7
23	0	1	2	2	3	4	4	5	6	7	7
24	0	1	2	3	3	4	5	6	6	7	8
25	0	1	2	3	3	4	5	6	6	7	8
26	0	1	2	3	4	4	5	6	7	8	8
27	0	1	2	3	4	5	5	6	7	8	9
28	0	1	2	3	4	5	6	7	7	8	9
29	0	1	2	3	4	5	6	7	8	9	9
30	0	1	2	3	4	5	6	7	8	9	10
31	0	1	2	3	4	5	6	7	8	9	10

Table B.5 Serial link drive strength configuration

Drive strength [1:0]	Iout mean (mA)	Vdiff (mV)	Description
00	1.95	438	Normal mode
01	1.61	348	Low power mode
10	2.87	610	High drive strength mode

VACFG[1] Raw mode Enables the sending of raw data samples. It is equivalent to running zero suppression with a threshold of 0, but it removes the two extra control words at the beginning of the data payload (time stamp and cluster size).

VACFG[2] Cluster sum Enables the cluster sum compression method.

VACFG[3] Huffman Enables Huffman compression.

VACFG[4] Send empty packets In case there are no pulses above the zero suppression threshold for a complete time window, which means there is no data to be transmitted in the payload of the packet, there will still be sent a header with an empty payload. If this is unwanted due to bandwidth restrictions or other causes, then this bit can be set to 0 to prevent them from being sent.

B.1.8 Power Saving Features

Some power saving features have been implemented. As power saving features sometimes might have unintended consequences which can render the design unusable, it is possible to disable them.

VACFG[5] Power save enable Setting this bit makes sure that the data pins of the pedestal memory are not toggling when the device is in a mode where the pedestal memory is not in use. When the pins are not toggling there will be no activity in the memory and it will consume less power.

VACFG[6] I^2C clock gate The state machine of the I^2C module can be automatically powered down when there is no activity on the I^2C lines to save power. This bit enables this feature.

VACFG[7] Neighbour block power down The neighbouring block, including its ring buffer and receiver circuit, can be disabled automatically to save power when the number of neighbours (NBCFG[7:6]) is set to 0, this feature enables this functionality.

B.1.9 Test Functionality

To aid in debugging and test of the chip, some extended functionality has been added.

ERRORS The packet headers are equipped with a Hamming error correction code which enables correction of one error and detection of two errors in the header. The Hamming code is added to the header before it is written to the ring buffer memory and it is checked again when it is read out of the ring buffer before it is sent to the serial links. The ERRORS register contains a counter for the number of errors found that could be corrected, and the number of errors that were not correctable and where the packet was dropped. The counters collect errors from all channel ring buffer outputs, in addition to the neighbouring ring buffer output and the neighbouring input.

CMD:ERCLR (Error clear) Clears the errors in the ERRORS register.

BYPASS The BYPASS register controls a multiplexer situated directly at the serialOut[0] output. See Table B.6 for specifics.

- **0x1-0x3** Feed-through from one of the trigger inputs can be used for determining round-trip delay of the system in a production environment. It can also be used for qualifying the SLVS driver/receiver pair.
- **0x4** Feed-through from the neighbour input DinN can be used for bypassing the device in the chain. This option also feeds NBflowstop_in to NBflowstop_out to enable full bypass of the device in daisy-chained mode.
- **0x5** Feed-through from delayed neighbour input after the neighbouring input has passed through the delay specified by NBCFG[5:0]. This can be used for qualifying the neighbour input delay chain. Also feeds NBflowstop_in to NBflowstop_out for full bypass in a daisy chain.
- **0x6** Linear Feedback Shift Register (LFSR) generator is intended for use in qualifying the SLVS driver. When selected it enables a 31 bit LFSR generator which feeds back the XNOR of output 31 and 28. It runs at the serial link clock.
- **0x7** The ADC test serializer selects the channel configured in the SERCHSEL register and serializes it at the serial link half clock with a preamble of b010011.
- **0x8-0xC** Internal clock output can be used for qualifying the quality of the clocks arriving from the clock divider. As there is a configurable delay chain (ADCDEL) between the digital ADC clock and analogue ADC clock it is possible to qualify the delay chain by providing an external ADC clock and outputting the analogue ADC clock through this BYPASS mode.
- **0xD** A ring oscillator has been implemented to determine process variations and thermal/voltage effects. The oscillation is only operative after the RINGOSCTST (Table B.3) command has been sent and then only for 256 ADC clock cycles or 512 oscillator cycles, whichever is the shortest. It should oscillate at about 100.9 MHz for the worst case, 220 MHz for best case and 160.9 MHz for the typical case.

SERCHSEL (Serializer channel select) When BYPASS is set to 0x7, then this selects the ADC channel used for the test serializer.

RINGCNT (Ring oscillator counter) In a non-test environment, it is possible to determine the process speed of a device by comparing the ring oscillator clock

Table B.6 Bypass signals for serial out 0 output

BYPASS[3:0]	Description
0x0	Serializer 0
0x1	Feed-through from bx_sync input
0x2	Feed-through from trg input
0x3	Feed-through from hb_trg input
0x4	Feed-through from neighbour input (dinN)
0x5	Feed-through from delayed neighbour input (dinN_del)
0x6	Output of 31 bits lfsr generator
0x7	Output of ADC test serializer
0x8	Internal ADC clock for digital part 10/20 MHz
0x9	Internal serial out clock divided by 2
0xA	Internal bunch crossing clock 40 MHz
0xB	Internal ADC clock for analogue part 10/20 MHz
0xC	Internal SAR ADC statemachine clock for analogue part 80/160 MHz
0xD	Clock from ring oscillator (only when triggered to run) 100–220 MHz

against the ADC clock. The ring oscillator clock is divided 16 times and is used to run a counter which is compared to a counter running at the ADC clock speed. The RINGCNT value is a signed byte and the oscillation period can be calculated as $OSC = \frac{255 \cdot ADC}{(255 - RINGCNT) \cdot 16}$ where ADC is the ADC clock period.

CMD:RINGOSCTST (Ring oscillator test) This command enables the above-mentioned ring oscillator test.

B.2 Channel Specific Registers

The registers listed in Tables B.7 and B.8 are specific to each channel. All channels can be addressed at the same time by using a broadcast command when writing. Refer to Sect. B.1.3 for information on how to access the channel registers.

B.2.1 Data Path Configuration

All the filters can be bypassed so that only the filters that suit the detector's needs are required to be enabled. Below are the specific configuration settings to set up the data path filtering.

Table B.7 Channel specific registers part 1

Register name		Address	Type	Default	Description	
K1	[12:0]	0x00	RW	0x000	[12:0]	First pole of the TCFU
K2	[12:0]	0x01	RW	0x000	[12:0]	Second pole of the TCFU
K3	[12:0]	0x02	RW	0x000	[12:0]	Third pole of the TCFU
K4	[12:0]	0x03	RW	0x000	[12:0]	Fourth pole of the TCFU
L1	[12:0]	0x04	RW	0x000	[12:0]	First zero of the TCFU
L2	[12:0]	0x05	RW	0x000	[12:0]	Second zero of the TCFU
L3	[12:0]	0x06	RW	0x000	[12:0]	Third zero of the TCFU
L4	[12:0]	0x07	RW	0x000	[12:0]	Fourth zero of the TCFU
L30	[12:0]	0x08	RW	0x000	[12:0]	TCFU IIR SOS first zero(L3) gain
ZSTHR	[11:0]	0x09	RW	0x000	[11:0]	Zero suppression threshold, 2 bit precision
ZSOFF	[12:0]	0x0A	RW	0x000	[12:0]	Offset added before truncation, 2's compliment, 2 bit precision
ZSCFG	[8:0]	0x0B	RW	0x000		Zero suppression configuration
			RW	0x000	[1:0]	Glitch filter, minimum accepted pulse, all, >1, >2, >2
			RW	0x000	[4:2]	Post-samples
			RW	0x000	[6:5]	Pre-samples
			RW	0x000	[7]	Change position of BC3 in pipeline (BC3 after BC2)
			RW	0x000	[8]	Enable Raw data output of ZSU
FPD	[12:0]	0x0C	RW	0x000	[12:0]	BC1 Fixed pedestal (offset subtracted), 2's compliment, 2 bit precision
VPD	[12:0]	0x0D	R	0x000	[12:0]	BC1 variable pedestal, 2's compliment
BC2BSL	[12:0]	0x0E	R	0x000	[12:0]	BC2 Computed Baseline, 2's compliment
BC3BSL	[12:0]	0x0F	R	0x000	[12:0]	BC3 Computed Baseline, 2's compliment
PMDATA	[9:0]	0x10	RW	0x000	[9:0]	Data to be stored or read from the pedestal memory
BC2LTHRREL	[9:0]	0x11	RW	0x003	[9:0]	BC2 lower relative threshold
BC2HTHRREL	[9:0]	0x12	RW	0x003	[9:0]	BC2 higher relative threshold
BC2LTHRBSL	[10:0]	0x13	RW	0x400	[10:0]	BC2 lower saturation level for baseline

(continued)

Table B.7 (continued)

Register name		Address	Type	Default	Description	
BC2HTHRBSL	[10:0]	0x14	RW	0x3FF	[10:0]	BC2 higher saturation level for baseline
BC2CFG	[10:0]	0x15	RW	0x000		BC2 configuration
			RW	0x000	[1:0]	Number of taps in moving average filter
			RW	0x000	[3:2]	BC2 pre-samples
			RW	0x000	[7:4]	BC2 post-samples
			RW	0x000	[8]	BC2 glitch removal
			RW	0x000	[10:9]	Auto reset configuration
BC2RSTVAL	[7:0]	0x16	RW	0x032	[7:0]	Reset value for maf baseline when auto reset is enabled
BC2RSTCNT	[7:0]	0x17	RW	0x0FF	[7:0]	Number of samples outside of thresholds before resetting maf filter (divided by 4)

ZSCFG (Zero suppression configuration)

7 Baseline correction order Enabling this moves the BC3 filter before the BC2 filter. The BC2 filter can then help reduce some of the noise in the baseline generated by the BC3 filter. The BC2 filter will then also not experience any large fluctuations in the baseline which could make it get stuck outside its thresholds and so it should be safe to use.

DPCFG (Datapath configuration)

3:0 BC1 mode Switches the data path in the Baseline Correction 1 filter between the different operation modes. See Table B.9 for the different modes.

6 Pedestal memory record Enables writing the data from the input directly into the pedestal memory. Data writing follows the time window and starts writing at address 0 from a trigger start and ends when the time window is completed.

7 DS enable Enables the Tail Cancellation filter correction.

8 BC2 enable Enables the Baseline Correction 2 moving average based filter correction.

9 BC3 enable Enables the Baseline Correction 3 slope based filter correction.

B.2.2 Digital Shaper

These are the configuration settings for the 3rd/4th IIR Tail Cancellation Filter/Data Shaper/generic IIR filter. See Sect. 3.2.3.2 for references to the poles and zeroes.

K1:K4 Poles of the IIR filter.

L1:L4 Zeros of the IIR filter.

Table B.8 Channel specific registers part 2

Register name		Address	Type	Default	Description	
DPCFG	[11:0]	0x18	RW	0x000		Data path configuration
			RW	0x000	[3:0]	BC1 mode, see Table B.9
			RW	0x000	[4]	BC1 data input polarity
			RW	0x000	[5]	BC1 pedestal memory polarity
			RW	0x000	[6]	BC1 pedestal memory record from input
			RW	0x000	[7]	TCFU enabled
			RW	0x000	[8]	BC2 moving average filter enable
			RW	0x000	[9]	BC3 filter enable
			RW	0x000	[10]	TCFU SOS architecture enable
			RW	0x000	[11]	TCFU signed Pole/Zero enable
BC1THRL	[10:0]	0x19	RW	0x7FD	[10:0]	Lower threshold of variable pedestal filter, 2's compliment
BC1THRH	[10:0]	0x1A	RW	0x003	[10:0]	Higher threshold of variable pedestal filter, 2's compliment
BC1CFG	[9:0]	0x1B	RW	0x114		BC1 configuration
			RW	0x004	[3:0]	Number of taps in variable pedestal filter
			RW	0x001	[4]	Define open threshold time of 31(high)/15(low) samples after (auto)reset
			RW	0x000	[5]	Force enable IIR also inside time window
			RW	0x000	[6]	High if BC1THR should be considered absolute, else relative
			RW	0x002	[8:7]	Shift output data of pedestal memory
			RW	0x000	[9]	BC1 negative clipping enabled
BC1RSTCNT	[7:0]	0x1C	RW	0x000	[7:0]	Number of samples outside of thresholds before resetting vpd filter (divided by 4)
			RW	0x000		0 disables the auto reset
BC3SLD	[7:0]	0x1D	RW	0x020	[7:0]	Rate of the BC3 baseline down counter
BC3SLU	[7:0]	0x1E	RW	0x010	[7:0]	Rate of the BC3 baseline up counter

L30 Extra pole included for generalizing the filter.

DPCFG (Datapath configuration)

- **10 Single order section architecture** When this is enabled the filter is changed from a 4th order filter to a 3rd order filter.
- **11 Signed coefficients** Enables signed Pole/Zero coefficients

B.2.3 Baseline Correction 1

These are the configuration settings for the various filtering capabilities of the first baseline correction filters.

FPD (Fixed pedestal) Each channel has a baseline value which usually is desirable to remove so that the signal is sitting at zero instead of at an offset from zero. The value is two's complement with two extra bits of precision.

VPD (Variable pedestal) The current baseline calculated by the variable pedestal filter.

DPCFG (Datapath configuration)

- **4 Invert input** Negates the input to the baseline correction 1 filter.
- **5 Invert output** Negates the output value of the pedestal memory.

BC1THRL (Threshold low) Sets the threshold below the current value of the VPD for when the VPD should not update its value. If BC1CFG[6] is enabled then this threshold is an absolute threshold instead of being relative to the current VPD value.

BC1THRH (Threshold high) Sets the threshold above the current value of the VPD for when the VPD should not update its value. If BC1CFG[6] is enabled then this threshold is an absolute threshold instead of being relative to the current VPD value.

BC1CFG (Baseline correction 1 configuration)

- **3:0 Taps** Sets the number of taps of the VPD filter.
- **4 Auto reset** When the signal has been outside the acceptance thresholds for a specified time given by BC1RSTCNT the filter will reset and open up the thresholds so that it can regain the correct baseline and avoid getting stuck. If this bit is set high it will open the thresholds for 31 cycles, and if it is low it will open them for 15 cycles.
- **5 VPD always on** The VPD filter can be set to only run when outside of a time window or it can be forced to run all the time when this setting is enabled. If running in continuous mode and one wants to use the VPD then this needs to be enabled.

Table B.9 Operating modes of the first Baseline Correction. The lookup function f( ) is the pedestal memory with the argument as the address

DPCFG[3:0]	Effect
0x0	din - FPD
0x1	din - f(t)
0x2	din - f(din)
0x3	din - f(din - VPD)
0x4	din - VPD - FPD
0x5	din - VPD - f(t)
0x6	din - VPD - f(din)
0x7	din - VPD - f(din - VPD)
0x8	f(din) - FPD
0x9	f(din - VFD) - FPD
0xA	f(t) - FPD
0xB	f(t) - f(t)
0xC	f(din) - f(din)
0xD	f(din - VPD) - f(din - VPD)
0xE	din - FPD
0xF	din - FPD

6 Absolute thresholds Changes BC1THRL and BC1THRH from being relative thresholds to being absolute thresholds.

8:7 Pedestal memory shift output Shifts the output value from the pedestal memory in the specified amount of steps. The final output value is 13 bits, two's complement with two-bit precision. With a value of 0, the pedestal memory value is 8 bits with 2 bits of extra precision, with a value of 1 the pedestal memory value is 9 bits with 1 bit of extra precision, with a value of 3 the pedestal memory value is 10 bits and with a value of 4 the pedestal memory value is multiplied by 2 and is 10 bits two's complement.

9 Clip output If this is set, then negative values on the output of the BC1 filter will be clipped to zero.

BC1RSTCFG (Reset configuration) Specifies the number of samples that the signal can be outside the thresholds before the auto-resetting steps in and the thresholds are opened. It is given in steps of four so the required number of samples must be divided by four. If the value is set to zero then the auto-resetting feature is disabled.

B.2.4 Baseline Correction 2

The following settings configure the various thresholds and filtering options for the second baseline correction filter.

BC2BSL (Baseline) The baseline as calculated by the BC2 filter.

BC2LTHRREL (Low threshold, relative) Sets the threshold below the current value of the calculated baseline for when the filter should not update its baseline value.

BC2HTHRREL (High threshold, relative) Sets the threshold above the current value of the calculated baseline for when the filter should not update its baseline value.

BC2LTHRBSL (Low threshold, baseline) Sets the minimum absolute value that the baseline can have before the auto-reset countdown starts.

BC2HTHRBSL (High threshold, baseline) Sets the maximum absolute value that the baseline can have before the auto-reset countdown starts.

BC2CFG (Baseline correction 2 configuration)

- **1:0 Number of taps** Select the number of taps in the moving average filter, 3 = 8 taps, 2 and 1 = 4 taps, 0 = 2 taps.
- **3:2 Pre-samples** Sets the number of samples before (pre-samples) the signal crosses out of the acceptance threshold that should be excluded from the averaging.
- **7:4 Post-samples** Sets the number of samples after (post-samples) the signal crosses into the acceptance threshold that should be excluded from the averaging.
- **8 Glitch filter** If this is enabled, then single samples that pass out of or into the thresholds do not trigger the pre- and post-samples.
- **10:9 Auto reset mode** Selects the method of auto-resetting the BC2 filter. A value of 3 selects the open threshold method of reset. Selecting 2, enable the auto-reset of the baseline to the current BC3 baseline, a value of 1 resets the baseline to the value given in BC2RSTVAL, 0 will disable the auto-resetting functionality.

BC2RSTVAL (Reset value) When BC2CFG[10:9] is set to 1, then this value is used when the BC2 filter needs to be reset (the value BC2RSTVAL is multiplied by 4 when loaded into a 13 bit signed data-path register).

BC2RSTCNT (Reset count) When auto-reset is enabled, then this specifies the number of samples outside the threshold before the auto-resetting steps in. It is given in steps of four so the required number of samples must be divided by four.

B.2.5 Baseline Correction 3

The below settings configure the gradients of the slope in the third baseline correction filter.

BC3BSL (Baseline) The baseline as calculated by the BC3 filter.

BC3SLD (Slope down) The rate at which the average changes in the downward direction when the signal is below the current average. The value is given with 0.25 bits precision.

BC3SLU (Slope up) The rate at which the average changes in the upward direction when the signal is above the current average. The value is given with 0.25 bits precision.

B.2.6 Zero Suppression

The following registers set the thresholds, offsets and other configuration options for running with zero suppression.

ZSTHR (Zero suppression threshold) Signals below this value will be suppressed and not included in the data stream. The value is given with 0.25 bits precision. Setting this value to zero effectively disables zero suppression as long as no filters are enabled which can bring the sample value below zero.

ZSOFF (Zero suppression offset) Since data is transmitted as 10 bits unsigned, there is a necessity to truncate the data to only positive values. To avoid losing information, like the tale of pulses that can pass below the average, it is possible to add an offset to all values before truncation. As this happens before the zero suppression threshold decision, any offset added here needs to be taken into account when setting the threshold.

ZSCFG (Zero suppression configuration)

1:0 A glitch filtering setting to remove spurious signals above the zero-suppression threshold. With this set to 4 , the pulses that are 3 samples or shorter will be removed. If it is 3, then pulses that are 2 samples or shorter will be removed. With this set to 2, then single samples above the threshold will be removed. Setting it to 0 will accept all pulses.

4:2 Samples before pulse starts Sets the number of samples before the signal passes above the threshold that should be included.

6:5 Samples after pulse ends Sets the number of samples after the signal has passed below the threshold that should also be included.

8 Raw samples Enables sending of samples that have not been modified by the BC2 or BC3 filter, but the zero suppression threshold decision is still done on the filtered baseline.

Appendix C
About the Author

Arild Velure has been working on electronics for particle detectors since 2010. He received his Bachelor of Engineering within the subject of electronics from Bergen University College in 2009 and his Masters in Physics within the subject of microelectronics from the University of Bergen in 2011. Velure was accepted for a Ph.D. stipend in spring 2013 at the University of Bergen. He was awarded his Ph.D. in Physics in spring 2019 for his work on the design of the SAMPA chip. He is now working as a fellow for the CERN ALICE Inner Tracker System developing systems for readout of pixel detectors.

C.1 Publications as Primary Author

1. Upgrades of the ALICE TPC front-end electronics for LS1 and LS2
   Velure, A. for the ALICE TPC Collaboration
   Proceedings of 19th IEEE-NPSS Real Time Conference (RT), 1–3, 26–30 May 2014
   https://doi.org/10.1109/RTC.2014.7097489
2. Upgrades of the ALICE TPC Front-End Electronics for Long Shutdown 1 and 2
   Velure, A. for the ALICE TPC Collaboration
   IEEE Transactions on Nuclear Science, Vol. 62, No. 3, 1040–1044, 2015
   https://doi.org/10.1109/TNS.2014.2382332
3. High speed continuous DAQ system for readout of the ALICE SAMPA ASIC
   Tambave, G. and Velure, A. for the ALICE Collaboration
   Proceedings of 20th IEEE-NPSS Real Time Conference (RT), 1–4, 6–10 June 2016
   https://doi.org/10.1109/RTC.2016.7543104

A. Velure, *A Digital Signal Processor for Particle Detectors*, Springer Theses,
https://doi.org/10.1007/978-3-030-71559-5

4. Qualification of the ALICE SAMPA ASIC with a High-Speed Continuous DAQ System
   Tambave, G. and Velure, A. for the ALICE Collaboration
   IEEE Transactions on Nuclear Science, Vol. 64, No. 6, 1461–1466, 2017
   https://doi.org/10.1109/TNS.2017.2707339

C.2 Publications as Co-Author

Run 3 upgrade

5. Upgrade of the ALICE Time Projection Chamber
   ALICE Collaboration
   ALICE Technical Design Report 16, 2014
   https://cds.cern.ch/record/1622286
6. Upgrade of the ALICE Readout & Trigger System
   ALICE Collaboration
   ALICE Technical Design Report 15, 2014
   https://cds.cern.ch/record/1603472

SAMPA

7. SAMPA chip: a new ASIC for the ALICE TPC and MCH upgrades
   Barboza, S.H.I. et al.
   Journal of Instrumentation, Vol. 11, No. 2, C02088, 2016
   https://doi.org/10.1088/1748-0221/11/02/C02088
8. Characterization of the first prototype of the ALICE SAMPA ASIC for LHC Run 3 and beyond
   Tambave, G et al.
   Journal of Instrumentation, Vol. 12, No. 3, C03012, 2017
   https://doi.org/10.1088/1748-0221/12/03/C03012
9. SAMPA Chip: the New 32 Channels ASIC for the ALICE TPC and MCH Upgrades
   Adolfsson, J et al.
   Journal of Instrumentation, Vol. 12, No. 4, C04008, 2017
   https://doi.org/10.1088/1748-0221/12/04/C04008
10. First irradiation test results of the ALICE SAMPA ASIC
    Mahmood, S.M et al.
    Proceedings Of Science, TWEPP-17 pp. 093, 2018
    https://doi.org/10.22323/1.313.0093

RCU2

11. RCU2 - The ALICE TPC readout electronics consolidation for Run2
    Alme, Johan et al.
    Journal of Instrumentation, Vol. 8, 2013
    https://doi.org/10.1088/1748-0221/8/12/C12032

12. First irradiation tests results of the ALICE TPC readout control unit 2
Zhao, Chengxin et al.
International Journal of Modern Physics A, Vol. 29, No. 24, 2015
https://doi.org/10.1088/1748-0221/10/01/C01016
13. Clock and data recovery methods for the readout control unit 2 in ALICE TPC
Torgersen, Christian et al.
Journal of Instrumentation, Vol. 10, 2015
https://doi.org/10.1088/1748-0221/10/04/C04028
14. Upgrade of the ALICE TPC FEE online radiation monitoring system
Røed, Ketil et al.
Journal of Instrumentation, Vol. 10, 2015
https://doi.org/10.1088/1748-0221/10/12/P12019
15. First performance results of the ALICE TPC Readout Control Unit 2
Zhao, Chengxin et al.
Journal of Instrumentation, Vol. 11, 2016
https://doi.org/10.1088/1748-0221/11/01/C01024

SRAM-based neutron detector

16. Design and characterization of an SRAM-based neutron detector for particle therapy
Ytre-Hauge, Kristian et al.
Nuclear Instruments and Methods in Physics Research Section A, Vol. 804, 64–71, 2015
https://doi.org/10.1016/j.nima.2015.09.049
17. First application of a novel SRAM-based neutron detector for proton therapy
Ytre-Hauge, Kristian et al.
Radiation Measurements, 2019
https://doi.org/10.1016/j.radmeas.2019.01.001

Additionally there are 205 publications as an ALICE collaboration member co-author.